Green Energy and Technology

More information about this series at http://www.springer.com/series/8059

Krishnan S. Hariharan · Piyush Tagade
Sanoop Ramachandran

Mathematical Modeling of Lithium Batteries

From Electrochemical Models to State Estimator Algorithms

Springer

Krishnan S. Hariharan
Samsung Advanced Institute of Technology (SAIT)-India
Samsung R&D Institute India
Bangalore, Karnataka
India

Sanoop Ramachandran
Samsung Advanced Institute of Technology (SAIT)-India
Samsung R&D Institute India
Bangalore, Karnataka
India

Piyush Tagade
Samsung Advanced Institute of Technology (SAIT)-India
Samsung R&D Institute India
Bangalore, Karnataka
India

ISSN 1865-3529 ISSN 1865-3537 (electronic)
Green Energy and Technology
ISBN 978-3-319-03526-0 ISBN 978-3-319-03527-7 (eBook)
https://doi.org/10.1007/978-3-319-03527-7

Library of Congress Control Number: 2017955822

Printed on acid-free paper

This Springer imprint is published by Springer Nature
The registered company is Springer International Publishing AG
The registered company address is: Gewerbestrasse 11, 6330 Cham, Switzerland

Dedicated to knowledge of all manifestations and admirers

Preface

An explosive market of Li-ion batteries has led to aggressive demand for fundamental understanding and mathematical modeling for accurate battery management systems (BMS). Researchers from multi-various backgrounds contribute from their respective background, leading to a lateral growth. Risk of this runaway situation is that indiscriminate use without appreciating the cohesive fundamentals often results in misinterpretation. It is worthy to note that the guiding principles are similar and it is the lack of clarity that impedes significant advancements. A pivotal contribution can hence be made by explaining the fundamentals in a coherent manner. Such an offering would enable researchers from multiple domains appreciate the bedrock principles and forward the frontiers. The contents of this book thus compliment the multitude of research publications in this domain by providing inner understanding.

Battery is an electrochemical system, and any level of understanding cannot ellipse this premise. The common thread that needs to run across from detailed electrochemical models to algorithms used for real-time estimation on a microchip is that it be physics based. Build on this theme, this book has three parts. Each part starts with developing a framework often invoking basic principles and ends with certain verified real-time applications. Part I deals with electrochemical modeling the second with model order reduction and the last with state and health estimation. The electrochemical model is derived from thermodynamics and a physics-based model order reduction is evolved. The state estimation is explained using a generic Bayesian framework and health estimation is described using Machine Learning (ML) tools. This book is unique to be the only one completely dedicated to battery modeling for all such components of BMS applications. We hope this offering would satisfy a practicing engineer with a basic perspective and a budding researcher with essential tools for a comprehensive understanding.

Bangalore, India
August 2017

Krishnan S. Hariharan
Piyush Tagade
Sanoop Ramachandran

Acknowledgements

This work is a compilation of research outputs of a dedicated team, past and present. The body of work consists of the research on the battery management systems (BMS) at the battery lab of Samsung Advanced Institute of Technology (SAIT)—India, Samsung R&D Institute India - Bangalore. Acknowledgements are due to Dr. Subramanya Mayya, Director, SAIT—India for support and encouragement, Samsung R and D Institute India—Bangalore. Foremost of the contributors is Senthil Kumar Vadivelu, the inventor of the reduced order model as presented in this work. Ashish Khandelwal contributed to the electrochemical thermal model and continues to be a vital source. Priya Gambhire was most productive during her short stay. Suman Basu is the SME in battery modeling and simulations. Aravinda, Rajkmar Patil, Mohan Verma, Meru Patil, Nandhini, and Shanthi are significant contributors to advancement in battery research that resulted in this work. Research at SAIT—India was initiated by the Energy Material center at SAIT-HQ, Korea. Authors acknowledge the center head and the energy lab leaders, and the active collaborators. One of the authors (KSH) would like to thank Dr. Prakash Bharathi, Director of the ersthwhile India Science Lab, Dr. Mark Verbrugge, Director at the General Motors R&D, Brian Koch, and the members of the VEC center. On a personal note, KSH would thank parents, Haripriya and Maitreya, for cruising on. SR would like to acknowledge the unwavering support from Neethu. PT would like to thank parents, siblings, and Neha for all the help and encouragement during the drafting of the book.

Acknowledgements are also due to all those who have helped us directly or indirectly in our efforts to make this book a possibility.

Contents

1 Introduction and Perspective 1
1.1 Use of a Book Without Pictures or Conversation? 1
1.2 Oh Dear! Oh Dear! I Shall be Late! 2
1.3 Down the Rabbit Hole 4
1.4 The Loveliest Garden 5
1.5 Caucus-Race—Best Way to Explain it is to do it! 7
1.6 Grin of the Cheshire Cat 8
1.7 Innovations in Each of the Chapters 9

Part I The Electrochemical Thermal Model

2 Theoretical Framework of the Electrochemical Model 13
2.1 Introduction 13
2.2 Conservation Equations 15
2.2.1 Mass Conservation in the Electrode (Solid) Regions 16
2.2.2 Mass and Charge Flux from Solution Thermodynamics . . . 16
2.2.3 Mass Conservation in the Electrolyte (Liquid Phase) 18
2.2.4 Charge Conservation in the Electrolyte (Liquid Phase). . . . 22
2.2.5 Charge Conservation in the Electrode (Solid) Region 25
2.3 The Charge Transfer Reaction 26
2.3.1 Nernst Equation: Equilibrium Component and its Thermodynamic Connect. 26
2.3.2 Butler–Volmer Equation: The Dynamic Component. 29

3 Theoretical Framework of Electrochemical–Thermal Model (ECT) 33
3.1 Energy Balance for Lithium Ion Cells: Lumped Heat Source 33
3.2 Energy Balance for Lithium Ion Cells: Distributed Heat Source 35

3.3 Basics of Nonequilibrium Thermodynamics 36
3.4 The Energy Conservation Equation with Distributed Heat Sources 38
3.5 The Coupled Electrochemical–Thermal Model 40
3.5.1 Thermal Model 40
3.5.2 The Electrochemical Sources of Heat Generation 40
3.5.3 Heat Generation Due to Phase Transition 41
3.5.4 Performance Indicators Computed from the Model 42

4 Key Applications of Electrochemical Theory 45
4.1 Introduction 45
4.2 Validation of Discharge and Charge 47
4.3 Analysis of Different Electrochemical Heat Sources 49
4.4 Effect of Ambient Temperature 50
4.5 Effect of Rate of Discharge 53
4.6 Effect of Heat Transfer Coefficient 55
4.7 Cumulative Heat Generation and Operating Efficiency Calculation 58
4.8 Summary 59

Part II Reduced Order Models

5 Theoretical Framework of the Reduced Order Models (ROM) 63
5.1 Introduction 63
5.2 Solid Phase Mass Balance Equations: A Single Sphere 66
5.2.1 The Diffusion Equation in Radial Coordinates and the Volume Averaged Version 66
5.2.2 Profile Approximations 67
5.2.3 Building the Battery Model 69
5.2.4 Butler–Volmer Kinetics 69
5.3 Solid Phase Current Balance Equations 71
5.3.1 Field Equations 71
5.3.2 Negative and Positive Electrode Regions 71
5.3.3 Revisiting the Battery Model 72
5.3.4 Insights from the Reduced Order Model 73
5.3.5 Electrode Potential Drop 74
5.3.6 Voltage Due to Electronic Conduction 75
5.4 Treatment of the Electrolyte Potential 76
5.4.1 Simple Linear and Nonlinear Approaches 76
5.4.2 Volume Averaging the Electrolyte Mass Balance Equation 77
5.5 Electrolyte Potential Field 86
5.5.1 Field Equations 86
5.5.2 Revisiting the Battery Model 90

5.6 Developments to the Base ROM 93
5.6.1 Thermal Balance 93
5.6.2 Coupling of Electrochemical and Thermal Models 96
5.6.3 Degradation: Cycling Studies 97
5.6.4 Phase Change 99
5.7 Summary of the ROM 101

6 Key Applications of ROM 105
6.1 Introduction 105
6.2 Representation of Experimental and Drive Cycle Data 106
6.2.1 Details of Experiments 106
6.2.2 Validation with CC Data 106
6.2.3 Representation of Drive Cycle Data 107
6.2.4 Time Taken 110
6.3 Parametric Sensitivity Analysis 111
6.3.1 Contribution to the Total Heat Release 111
6.3.2 Effect of Ambient Temperature 112
6.3.3 Effect of Rate 114
6.3.4 Effect of Heat Transfer Coefficient 116
6.4 Representation of Phase Change Electrodes 118
6.4.1 Reduced Order Model for Phase Change System 118
6.4.2 Reduced Order Model for Phase Change System with Degradation 119
6.4.3 Charge–Discharge Asymmetry 121
6.4.4 Path Dependence 122
6.5 Summary 124

Part III State and Health Estimation

7 Theoretical Framework for State Estimation 129
7.1 Introduction 129
7.2 Bayes Theorem 130
7.3 Recursive Bayesian Framework 133
7.3.1 Kalman Filter 136
7.3.2 Extended Kalman Filter 138
7.3.3 Unscented Kalman Filter 140
7.3.4 Particle Filter 141

8 Theoretical Framework for Health Estimation Using Machine Learning 145
8.1 Introduction 145
8.2 Primer on Machine Learning Algorithms 146
8.2.1 Machine Learning for Regression 147
8.2.2 Machine Learning for Classification 149
8.2.3 Gradient Descent Algorithm 150
8.3 Artificial Neural Network 154

8.4 Support Vector Machine . . . 158
8.5 Gaussian Process Emulator . . . 160
8.6 Recent Advances: Deep Learning Algorithms . . . 165

9 Key Applications of State and Health Estimation . . . 175
9.1 Li-Ion Battery State Estimation Using the Recursive Bayesian Filter . . . 176
9.2 State of Health Estimation Using Machine Learning . . . 190
9.2.1 Case 1: Multiple Battery Data . . . 192
9.2.2 Prognostic Horizon of Multistep Approach . . . 197
9.2.3 Case 2: Single Battery Data . . . 198
9.2.4 Case 3: Two Batteries at Various Temperatures . . . 200
9.2.5 Case 4, 5: Multi-battery Data at High and Low Temperatures (HT/LT) . . . 200
9.2.6 Case 6: All Battery Data . . . 202
9.3 Summary . . . 203

References . . . 205

Index . . . 209

Chapter 1
Introduction and Perspective

In this introductory chapter, the need for a comprehensive and in-depth understanding of the underlying physics of batteries is discussed. The layout of the book as well as the major themes are briefly introduced to set the tone for the rest of the chapters that follow.

1.1 Use of a Book Without Pictures or Conversation?

Over the last couple of decades, our society has witnessed a tremendous growth in mobility of devices. This is a paradigm shift from an earlier scenario, where the devices were predominantly stationary, and we would have to reach them to use them. To keep pace with the mobility of our society, desktop computers have given way to laptops, desk phones to mobiles, and so on. This shift has resulted in a multifold increase in demand for portable energy sources, and with it, the importance of batteries.

The subsequent era of miniaturization yet power hungry applications required higher amounts of energy be packed in lesser volume. A natural choice of the active material was lithium—being the metal with the lightest weight, highest voltage and hence the highest energy density. At present, lithium-ion batteries have found widespread applications in ranging from mobile phones to electric vehicles. In short, in almost all consumer electronic, power electronics, communication as well as transportation domains.

Lithium-ion batteries fall under the category of secondary batteries. These are batteries that can be recharged once it is completely discharged allowing for repeated usage. This rechargeable feature enables the lithium-ion batteries to be used as an onboard energy storage source in various industrial applications—be it a mobile phone, camera, or an electric vehicle. Although lithium-ion battery is not the first of its kind in being a rechargeable battery, it is far ahead of many of its competitors

K. S. Hariharan et al., *Mathematical Modeling of Lithium Batteries*,
Green Energy and Technology, https://doi.org/10.1007/978-3-319-03527-7_1

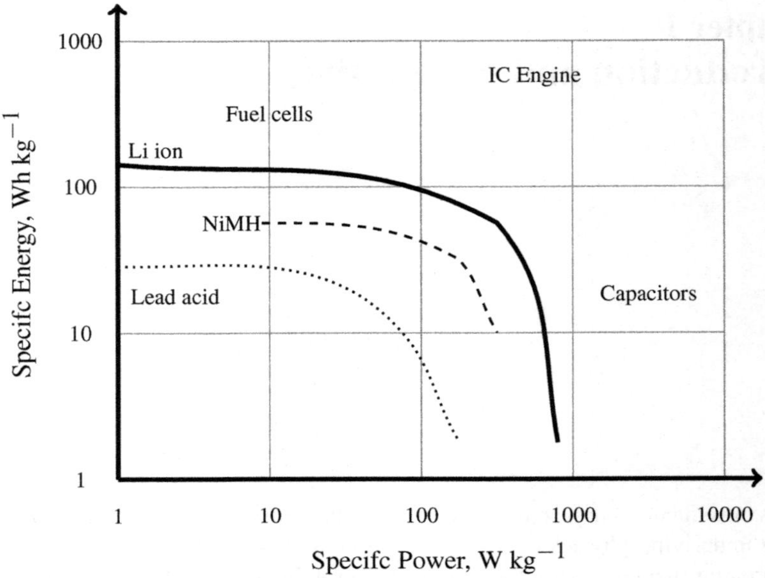

Fig. 1.1 Ragone plot (for illustration only)

at present due to many favorable features—with some inevitable unfavorable side effects. One of the features in favor of a lithium-ion battery is that it can store larger amounts of energy per unit mass/volume, i.e., have a higher energy density than many other cells. Any comparison chart between various cells would show lithium-ion cells at the top of the list. The corollary to this feature is that any given energy target can be achieved by lesser number of cells.

A typical "Ragone plot" shown in Fig. 1.1 gives the specific energy versus power comparison that can be delivered by battery types is schematically illustrated. Power pertains to acceleration, how quickly can one get the stuff out, and the energy pertains to the range—how long will the battery last. Lithium-ion cells are the clear winner, in addition to energy density, terms of higher discharge voltage, less self-discharge, etc. These advantages come with an equal share of disadvantages also, the singular issue being the heat release and the related safety issues. Cost, localized lithium resources, and lack of a clear method of recycle are some other aspects adding to further questions on the longevity of this technology.

1.2 Oh Dear! Oh Dear! I Shall be Late!

The deeper reason for such high energy densities is due to faster chemical reactions, or rather a lack of the slower ones in lithium-ion cells. Lithium-ion cells, unlike other secondary batteries, do not have any redox chemical reactions, hence there are no new chemical products that are formed. During operation, the lithium gets in and out of the electrodes by an intercalation reaction. The electrodes have open structures into which the lithium inserts itself, and is subsequently extracted from the electrodes as the battery undergoes discharge. In normal conditions, there are no unwanted side reactions, and hence the operation is almost completely reversible between charge and discharge.

The basics of electrochemistry as well as the advantages and drawbacks are shared by most of the lithium-ion cells. At a deeper level of granularity, however, the different chemistries differ in the type of electrode used—the positive electrode to a larger extent, as most industrial application use some form of graphite as the negative electrode. Based on the different crystal structure of the electrode material, the way in which lithium is stored in the positive electrode differs. On taking in lithium various electrodes attain different cell voltages. Diving deeper into the electrochemistry, depending on the way the electrode material responds to lithium interaction there are some exotic features exhibited like phase change, resulting in additional processes. The search of the ideal electrode is a high thrust area in lithium-ion battery research. The intermediate goal seems to be the ideal positive electrode, given graphite as the negative electrode. In order to combine the advantages of individual chemistries, the emergent direction of research is focused on mixing individual compounds to result in composite cathodes.

The purpose of identifying the chemistry aims at coming up with the optimum cell that meets the energy and power of the targeted application. For devices with lesser real estate like a mobile phone, only single cells are typically used. To be used as the energy source in automobile many of the cells are stacked up in a series–parallel combination to meet the energy requirements of the vehicle. The principal source of energy being the chemical energy released during the intercalation reaction mentioned earlier. Thus the individual cells are fabricated in various shapes and sizes be it a pouch, prism or a cylinder, assembled to battery packs and with the suitable electronic circuit support, is ready to power the electric vehicle. Onboard the device, the singular key indicator of the performance of the battery pack is the state of charge (SOC). Simply put, the SOC is the fuel gauge of the device, indicating how much usable charge is available in the battery pack. As the only charge carriers in a lithium-ion battery are the lithium stored in the electrodes itself, the measurement of the available charge pertains to the amount of lithium available in the electrodes. Unfortunately, this is not a directly measurable quantity unlike the fuel guage in a conventional internal combustion engine vehicle. It is at this point that the employment of a mathematical model becomes imperative, and the state of the battery pack is estimated (rather than measured) from a measurable signal.

1.3 Down the Rabbit Hole

To appreciate this point better, one needs to look at the battery discharge in detail. Discharge starts with all lithium in the negative electrode, say, graphite. Lithium that is inserted within the graphite particles is released, diffuses through the particle, and reaches the surface of the particle. At the surface of the active material particle at the electrode–electrolyte interface, neutral lithium releases an electron, and becomes ionized and enters the electrolyte. The electron, meanwhile, is conducted through the electrode travels through the external circuit and reaches the positive electrode. At the same time, diffusing through the electrolyte the lithium ion reaches the positive electrode, reacts with available electron and the process of reaction–diffusion repeats in the positive electrode. Thus, during discharge, lithium-ion concentration in both the electrodes varies as time proceeds. In addition, there is a spatial variation of lithium concentration across the thickness of the cell.

The voltage attained by the cell just due to the presence of lithium in the electrodes in the absence of any of the transport processes is the open-circuit voltage (OCV). All the transport processes involve a particular voltage drop and the cell voltage is lower than the OCV during discharge. During charging, however, the charge voltage has to be higher than the OCV to account for all the transport processes. Eventually, battery is said to be completely discharged when all the available lithium from the negative electrode is transferred to the positive electrode. The processes described are an ideal scenario with no limiting steps. If the rate of discharge were too high in comparison to rate of diffusion, discharge would be cut short earlier and not all lithium can be transferred. This is because a high rate of reaction with low rate of diffusion within the electrode particles can result in an accumulation lithium ions near the surface resulting in saturation. These processes will be discussed further when the electrochemical model for batteries is described.

Thus, a simple discharge involves a concerted occurrence of various electrochemical processes, such as

- diffusion in the solid phase,
- charge transfer reaction,
- electronic conduction,
- diffusion in the electrolyte phase,
- charge transfer reaction and
- diffusion in the solid phase.

An efficient utilization of the battery involves tuning the rates of the individual processes to be of comparable magnitudes. The rates being dependent on the amount of concentration of lithium in the electrodes, in turn maps to efficient cell design in terms of optimum thickness of the electrodes.

Given that a normal operation of a lithium-ion cell involves various processes, it can be seen that the coordination amongst these processes is more involved once the timescales of these processes are analyzed. Electronic conduction in typical electrode materials is quick, although there are exceptions, and happens in milliseconds. The charge transfer reaction is next and occurs in less than a second. Conduction in the electrolyte phase can span multiple timescales, and solid phase diffusion happens in 10–100 s. There are of course processes slower than these, for example, the interphase movement in electrodes that exhibit multiphase coexistence, etc. Thus, it can be seen that the timescales of the processes span a wide range of time as well—from 1 ms to 100–1000 s. The timescale analysis becomes the obvious tool for characterization tools like the electrochemical impedance spectroscopy (EIS).

Coupled with the timescales of the electrochemical processes is the length scales involved in a lithium-ion cell. Staring from the smallest length scale, particles in an electrode are of micrometer in radius. The individual electrode thickness is also of microns, so is the separator region that includes the electrolyte. Thus, a single electrochemical unit of positive and negative electrode and the electrolyte is of the order of 100 of microns. This is not a wide enough range, but a single electrode pair cannot generate enough power for any practical application. Many such unit electrochemical cells are stacked together to give a pouch cell or wound to give a cylindrical cell. Such a pouch cell can give a voltage in the range of 3–4 V and current of around 15 A, a cylindrical cell around 2–3 A. The cell used in mobile phones is of centimeters in dimension, and is enough to charge a mobile phone, but not an automobile. An automobile requires power of around 120 kW for which multiple single cells are stacked together to make a module, and many such modules, a battery pack. A battery pack can contain around hundreds to thousands of cells.

1.4 The Loveliest Garden

We started off this discussion to solve the problem of finding the key indicator of battery capacity, the SOC. This is still an open research as well as an engineering problem. Typically, a problem is unsolved due to lack of understanding of the fundamental physics, or due to weak mathematical framework. SOC estimation remains unsolved due to an additional and completely different reason that makes this an exciting area of research. The quantity that we are after the SOC is not a measurable quantity unlike temperature or concentration. It is by definition, an estimated quantity making it open for many approaches. Therein lies a more fundamental problem. SOC as an estimated quantity is not clearly defined either. This is more or less like entering the forest with a specific weapon, but unsure of the game.

In the state of the art, SOC is computed by Coulomb counting. This pertains to measuring the current at every time step, and integrating over time. This quantity is the net capacity the cell has lost during the particular operation and by subtracting from the total capacity, the available capacity or SOC is estimated. With an analogs scenario of a water tank, this method is equivalent to measuring the water outflow over a period of time. Knowing the total water the tank can hold, what is left in the tank can be inferred. Although intuitive and simplistic, it suffers from some important drawbacks. First, the initial capacity is known only if the cell is known to be in a fully charged or discharged state. Once the battery is taken through a cyclic operation, the initial capacity may not be known accurately. Additionally, many conditions like temperature affect the maximum cell capacity. Thus, if the cell undergoes a non-isothermal operation, the maximum capacity (the total water the tank can hold) keeps shifting making this method completely erroneous. Current sensors can be inaccurate, and operational issues such as sensitivity of measurement errors make this estimation method unreliable. An alternate method is to compute from the cell voltage itself. Cell voltage is a result of all the processes that has been discussed till now. Some of them like electronic conduction, do not necessarily result in net capacity change. Thus, during abrupt transients, the voltage based methods are unreliable.

An alternate method stems from the fact that the OCV of an electrode is a unique function of the SOC. Electrochemically speaking, the OCV of an electrode is a function of the lithium concentration in the electrode. This information is used in a cell level where the electrode OCV is replaced with cell OCV and concentration with the SOC of the cell. Here, the latter is typically an estimate from Coulomb counting. So the task is to compute the cell OCV and using known OCV-SOC relationship to reverse engineer the SOC. OCV is defined as the cell voltage at equilibrium. But while being used on board, the system is far from equilibrium. Thus, a mathematical model that gives out the OCV from measured voltage and current signals becomes imperative. A model obviously involves parameters and requires information at some granularity of the transport processes. To obtain OCV from the cell voltage effect of all transport processes have to be removed. This process involves some method of model inversion and comes with the usual complications.

An approach that is getting much attention recently is that of using electrochemical models for state estimation. Electrochemical models as is would be difficult to use due to the complicated mathematical structure. Hence, some type of model order reduction is sought. Using this approach, one could potentially compute the lithium concentrations within the electrodes thus bringing a paradigm shift in state estimation. With Coloumb counting, the state was inferred by what comes out of a cell. By estimating concentrations what exactly is within the cell can be known. The methodologies are not standard and are an active area of research. As discussed earlier, during the battery operation, the lithium concentration changes with time and space. Hence, an order reduction can be achieved by removing the spatial dependence by some neat trick. Retaining the physical basis, one could integrate out the space component and recast the model using time-varying volume averaged quantities. Alternatively one could explore other numerical ways of reduction using spectral projection in an orthogonal basis set.

The key performance indexes for any battery management system (BMS) are: (1) state, (2) health, and (3) safety. The state pertains to how long the battery will under the given conditions, health pertains to how long it would take to replace the battery and most importantly but often neglected is safety. This book deals mostly with the first two aspects.

An aggressive power demand translates to abrupt transients in the current flow, resulting of overlap of the clean charge–discharge processes described earlier. Hence to obtain the correct internal variables such as the lithium concentrations in the electrodes and electrolyte is an unwieldy task. Integrated with the model, a tool for state estimation is required. State estimation is a well-established field, and is customarily done using various filters the celebrated Kalman filter being most easy to use. Overuse without a clear perception of the basics brings the risk that state estimation has denigrated to curve fitting. In retrospect, this singular problem haunts the whole of battery modeling and is one of our main objectives is to eliminate the ambiguities.

Accurate estimation of the state tells us how long the battery would last till the next recharge. An associated issue the life of the battery itself which tells us when would it need to be replaced. Estimation of life is connected to successive charge–discharge cycles, and happens over hundreds of cycles. The effects of capacity fade can be felt however even during normal usage, as the duration between successive charges decreases as the device is used. The capacity of a lithium-ion battery fades due to undesired chemical reactions. These reactions result in irreversible loss of lithium that is trapped in the product of the chemical reactions. Predominantly ascribed to reaction between electrolyte and graphite electrode, recent research indicates that the positive electrode can also contribute to capacity fade. We could say that the state estimation solves an acute problem while the health estimation deals with a chronic one! The state of the art for health estimation is to use simple correlations that are limited in scope and insight. To make accurate predictions, the complete relevant history needs to be processed as a whole. With the advent of data analytics and machine learning, novel methods are emerging that are capable of elucidating meaningful correlations.

1.5 Caucus-Race—Best Way to Explain it is to do it!

An explosive market requirement has resulted in aggressive pursuits in the domain of battery research. Various researchers from multi-various backgrounds contribute based on their respective backgrounds, leading to lateral growth. The risk this run-away situation results is that researchers tend to use an existing method or algorithm, without an in-depth knowledge of the fundamentals—often misinterpreting the results. It is worthy to note that the guiding principles however, are similar, and the lack of clarity is what impedes further advancement. A repeat or even a synopsis of all the results and applications of the battery models and algorithms would be a mammoth task, and cannot be accomplished in any one offering. Hence, we believe that a pivotal contribution can be made by explaining the fundamentals in a simple coherent manner. Such an offering would enable the researchers from multiple domains to appreciate the bedrock principles and take it forward.

Having observed that the electrochemical models as is cannot be used on board, most of the industrial applications use equivalent circuit models for state and health estimation. Although intuitive, over-abstraction of the intricate electrochemical processes as resistors and capacitors results in frequent parameter estimation as a penalty. As the parameters are not mapped uniquely to the intrinsic processes this approach, by construct, has no predictability. An alternate route, popular of late, that retains the physical premise of the electrochemical model and enables onboard applications is to find various numerical methods for partially solving the electrochemical model. Various reduced order models have been reported in this regard, with increasing levels of complexity. It is often observed that the intricate numerical machinery clouds over the physical insights—and the output, in terms of the physics, is no better than the equivalent circuit models.

1.6 Grin of the Cheshire Cat

Battery is an electrochemical system, and any understanding of this system cannot eclipse this basic premise. Hence, the common thread that runs through all the models and algorithms described in various chapters of this work are physics based. Based on this premise, the manuscript has three parts. The first part deals with the detailed electrochemical modeling. Thus, the first chapter of this part derives the basic electrochemical model from thermodynamics and transport phenomena. The thermal models are integrated into the framework in the subsequent chapter. The last chapter in this part discusses some salient as well as representative results.

The second part discusses evolution of a reduced order model. Model order reduction using volume averaging is then explained—which is central to later developments. Application of this model to realistic data is discussed in the last chapter of this part.

The third part discusses state and health estimation. Integration of state estimation into the reduced order model, with the ensuing physical insights is described in the first chapter of this part. The health estimation, pursued using machine learning tools is discussed next. In this chapter, shallow learning and deep learning methods are discussed. A unique contribution of our work is the inclusion of fundamental derivations of machine learning (ML) algorithms. Specific mention of most recent deep learning tools like the Restricted Boltzmann Machine (RBM) and Deep Belief Network (DBN) is made. The final chapter in this part discusses about the results and physical insights on state and health estimation.

To summarize, the first part has the complete physics-based description of the lithium-ion battery. A reduced model with some compromise on the framework, but not the insight is the theme of the second part. First portion of the third part is the hybrid of physics and data-driven approaches, evolving to a completely data-driven approach in the last part. Within each part, the first or the first couple of chapters are on the theory and rigorous derivation. In the last chapter of each part, we have compiled certain key results as a demonstration. The contents of these respective chapters are an outcome of the various publications from our group, and have been referred to appropriately.

We hope this offering would satisfy a practicing engineer with a fundamental perspective, and a budding researcher with the essential tools required to develop a comprehensive understanding of the BMS development domain to advance the frontiers. It should be noted, however, that the publications dwell more on the applications of the modeling frameworks, and the derivations are not discussed in them. This work compliments the publications and provides more fundamental insights that can be developed to further research domains.

1.7 Innovations in Each of the Chapters

- Part I. Chapter 2: Fundamental components of the electrochemical model for Li-ion cell are derived from flux–force relations and basic thermodynamics.
- Part I. Chapter 3: The energy balance for the Li-ion cell derived from governing principles of nonequilibrium thermodynamics.
- Part I. Chapter 4: Application of the electrochemical thermal model.
- For Part I, also refer *Journal of Power Sources 248 (2014) 101–114* [28] and *Journal of Power Sources 279 (2015) 180–196* [27].
- Part II. Chapter 5: Physics-based order reduction of the electrochemical thermal model with intuitive physically relevant variables.
- Part II. Chapter 6: Application of reduced order model to realistic data.
- For Part II, also refer *Journal of Power Sources, 222 (2013) 426–441* [47], *Journal of Power Sources 248 (2014) 383–387* [48], *Journal of Power Sources 270 (2014) 281–291*, [15] *Journal of Power Sources 290 (2015) 87–101* [16].
- Part III. Chapter 7: State estimation developed using nonlinear filters; as special cases of a generic Bayesian estimation framework.
- Part III. Chapter 8: Novel machine learning algorithms are provided by fundamental derivation for the learning rules.
- Part III. Chapter 9: Application of state and health estimation.
- For Part III, also refer *Journal of Power Sources 306 (2016) 274–288* [52], *274–288, Applied Energy 159 (2015) 285–297* [36].

Part I
The Electrochemical Thermal Model

Chapter 2
Theoretical Framework of the Electrochemical Model

The basic principles of the electrochemical model for Li ion battery is developed from fundamentals of thermodynamics and transport phenomena. The evolution of the electrochemical model and the inherent assumptions are discussed. The discussions and derivations are self-consistent and complete. Mathematical model for each process in the Li-ion cell is constructed in a stepwise manner to evolve the complete electrochemical model.

2.1 Introduction

The framework for modeling lithium-ion batteries from a physics perspective essentially involves the specification of the mass and charge conservation equations in the various regions of the battery at a sufficiently coarse scale.

The basic components of the lithium-ion cell are the solid positive and the negative electrodes sandwiching a separator which is an electron insulating material. The whole region of the battery is then filled with a liquid electrolyte (a majority of the commercial cells still use liquid electrolytes). As a matter of terminology, active components refer to those that participate in the essential physicochemical processes that convert the chemical energy to electrical energy and vice versa. Thus, the fillers or binding material are not directly considered.

The electrodes constituting the battery are porous in nature. This is an artifact of the synthesis process which grossly involves mixing of the active material in a solvent and then allowing the solvent to dry. Depending on the intensity of the crushing of the material as well as the packing density, the porous structures can have a pore size distribution typically of the order of a micron. These porous structures enhance the contact surface area of the active material with the electrolyte and increase the probability of the charge transfer reaction.

K. S. Hariharan et al., *Mathematical Modeling of Lithium Batteries*,
Green Energy and Technology, https://doi.org/10.1007/978-3-319-03527-7_2

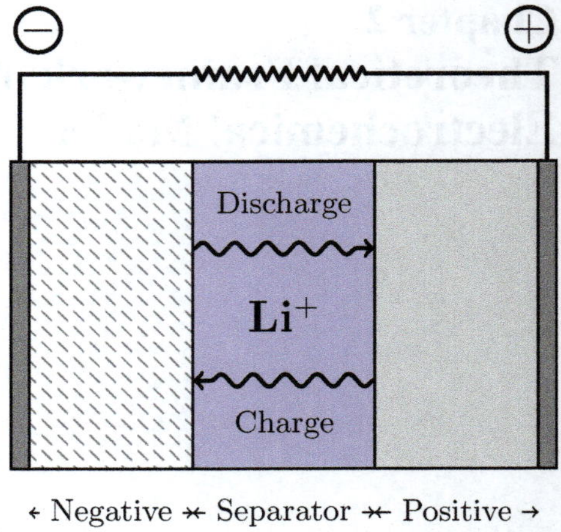

Fig. 2.1 Schematic diagram of the lithium-ion battery. During charge, the lithium ions move from the positive to the negative electrode. During discharge, the inverse direction is followed

This allows the basic framework in which the electrodes are modeled by appealing to the established porous electrode theory. Reference [32] The basic picture of an electrode is a porous solid filled with electrolyte in its interstices.

At the level of the length scale of the porous electrode theory, however, the differentiation between a pore and the solid is not made. This allows both the electrode as well as the electrolyte phase to be modeled as continuous media, implying that both the phases are assumed to exist at all points in space. Additionally, any point in space is considered electrically neutral. Thus, the model treats the system at a length scale larger than individual charge carriers, and smaller than the electrode dimensions. A schematic of a lithium-ion cells is given in Fig. 2.1.

The principal process during discharge is the transfer of lithium ion from the positive electrode to the negative electrode through the electrolyte, and in the reverse direction during charge. Each individual electrode section consists of the solid phase consisting of the active material, and the electrolyte, through which the lithium is transferred. The combined picture of a cell has a continuous electrolyte medium, interposed with solid electrode particles at the respective electrode sections. As a cell is a closed system, mass conservation is applied to the lithium that exists in the ionized state in the electrolyte. As the lithium ion carries a unit charge with it, charge is also simultaneously conserved. Similar to the electrolyte, these conservation laws also apply to the electrode. The interaction between the solid particles of the electrode and the liquid electrolyte happen at the interface, where charge is transferred. This charge transfer reaction couples the electrode and the electrolyte phases.

From the above description, it can be seen that any given electrode has to satisfy a couple of conservation laws—one for the mass of lithium and the other for the charge—in each of the solid and the electrolyte phases, with the charge transfer reaction coupling both of them. Each electrode region can, therefore, be represented by a total of five equations, and the separator region by two equations as this region is characterized by the single electrolyte phase with no charge transfer reaction.

This set of 12 equations forms the core of the electrochemical model or the macro-homogeneous model developed based on the continuum picture for the cell.

The mathematical model involves several levels of abstraction namely:

1. The solution methodology represents the porous solid phase as a collection of particles that are individually surrounded by the electrolyte phase.
2. The framework represents the electrode as a homogeneous system of electrolyte and particles at every point.
3. Due to the first abstraction, the diffusion of lithium in solid phase between particles is neglected. Diffusion in the solid phase is modeled within each particles that are typically assumed to be spheres.

The diffusion equation, which is the mass conservation equation in the solid phase, is solved in spherical coordinates to obtain the lithium surface concentration in the spheres. As the solid phase is also continuous, the surface concentration is obtained at every point in the electrode region. The surface concentration is the driving force for charge transfer reaction. The reaction flux, in turn, appears as the boundary condition for the solid phase diffusion equation. Thus, the model results in a set of equations that are fully coupled.

It is to be noted that all the equations in the model are solved as a function of the thickness of the electrode, except the diffusion in the solid phase which is solved in the spherical coordinates. Due to this effective mapping of the output of the spherical diffusion equation onto other field equations that are solved in Cartesian coordinates, this model is also called as a Pseudo-2D (P2D) model [32].

2.2 Conservation Equations

The governing equations of the P2D model can be derived by appealing to the charge and mass conservation equations in the various regions of the battery. These are derived in the following subsections. The dimensions of the LiB model electrodes and separator are illustrated in Fig. 2.2. The thickness of positive electrode, separator, and negative electrode are given by L_p, L_s and L_n, respectively.

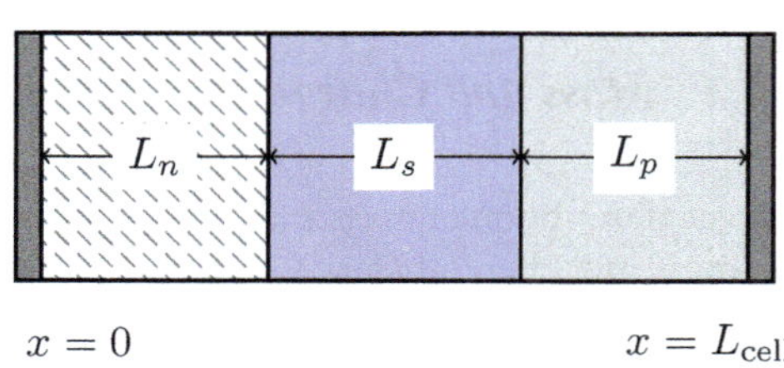

Fig. 2.2 Dimensions of the battery model components

2.2.1 Mass Conservation in the Electrode (Solid) Regions

In this section, the mass conservation equation in the solid phase of the active materials for both the electrodes is derived. The porous electrode constitutes of individual particles represented as spheres—that interact through the electrolyte that surrounds each particle. Lithium diffuses into the active material particles after gaining an electron by the charge transfer reaction at the electrode–electrolyte interface, it is considered as neutral lithium. During the charging or discharging processes, the lithium ion undergoes diffusion inside the solid porous electrodes. The porous electrode region is abstracted to be composed of individual particles, assumed to be spheres of finite radius, that interact with the electrolyte that surrounds each particle.

Lithium ion undergoes the charge transfer reaction at the electrode–electrolyte interface, becomes neutral by gaining an electron, and diffuses into the active material particles Hence, the mass conservation in the solid phase, in the most simplistic representation, is provided by the Fick's law of diffusion in the spherical coordinates. This continuity equation is given by,

$$\frac{\partial c_{si}}{\partial t} = \frac{1}{r^2}\frac{\partial}{\partial r}\left(D_{si} r^2 \frac{\partial c_{si}}{\partial r}\right), \tag{2.1}$$

where c_{si} is the concentration of lithium in the solid particles, with $i = p, n$ providing an index for either the positive or negative electrode. The spherical symmetry results in the boundary condition at the center of the sphere

$$\left.\frac{\partial c_{si}}{\partial r}\right|_{r=0} = 0. \tag{2.2}$$

The concentrations of the solid and the electrolyte phases are connected by the condition at the surface of the particles. At the surface, the flux of lithium is given by the pore wall flux, given by the charge transfer reaction.

$$-D_s \left.\frac{\partial c_s}{\partial r}\right|_{r=R_{pi}} = j_i. \tag{2.3}$$

2.2.2 Mass and Charge Flux from Solution Thermodynamics

A basic description of mass and charge flux in the solution phase is required to derive the mass and charge balance equations in the electrolyte region.The framework is derived in this subsection. Derivations for the mass and charge conservation equations resulting in the electrochemical model for Li-ion battery are presented in the next subsections.

Based on the concentrated solution theory, the flux is defined as

$$\mathbf{N}_i = -\mathscr{M}_i c_i \nabla \mu_i, \tag{2.4}$$

where $\mathscr{M}_i$ is the mobility of ion i, and is given by the Einstein relation to be

$$\mathscr{M}_i = \frac{D_i}{k_B T}, \tag{2.5}$$

where D_i is the mass diffusivity, k_B the Boltzmann constant, and T is the absolute temperature. Corresponding to the mass flux, a current flux can be defined as

$$\mathbf{J}_i = z_i e \mathbf{N}_i. \tag{2.6}$$

For electrochemical systems, μ_i is constructed including the contribution from the activity a_i , charge on the ion z_i and the electronic charge e, as well as the potential ϕ, as given below

$$\mu_i = k_B T \ln(a_i) + z_i e \phi \tag{2.7}$$

Activity is related to concentration c_i through the activity coefficient γ_i via the relation $a_i = c_i \gamma_i$. Thus,

$$\mu_i = k_B T(\ln(c_i) + \ln(\gamma_i)) + z_i e \phi \tag{2.8}$$

and

$$\nabla \mu_i = k_B T \nabla \ln(c_i) + k_B T \nabla \ln(\gamma_i) + z_i e \nabla \phi \tag{2.9}$$

$$\nabla \mu_i = k_B T \frac{\nabla c_i}{c_i} \left[1 + \frac{\nabla \ln(\gamma_i)}{\nabla \ln(c_i)} \right] + z_i e \nabla \phi = k_B T \frac{\nabla c_i}{c_i} \left[1 + \frac{\partial \ln(\gamma_i)}{\partial \ln(c_i)} \right] + z_i e \nabla \phi. \tag{2.10}$$

Based on this, the flux is written as

$$\mathbf{N}_i = -\mathscr{M}_i c_i \nabla \mu_i = -\frac{D_i}{k_B T} c_i \left(k_B T \frac{\nabla c_i}{c_i} \left[1 + \frac{\partial \ln(\gamma_i)}{\partial \ln(c_i)} \right] + z_i e \nabla \phi \right) \tag{2.11}$$

On simplification,

$$\mathbf{N}_i = -D_i \left[1 + \frac{\partial \ln(\gamma_i)}{\partial \ln(c_i)} \right] \nabla c_i - \frac{D_i}{k_B T} c_i z_i e \nabla \phi = D_I \nabla c_i - \frac{z_i e}{k_B T} D_i c_i \nabla \phi \tag{2.12}$$

The electric conductivity σ_i is defined as

$$\sigma_i = (z_i e)^2 \mathscr{M}_i c_i = (z_i e)^2 c_i \frac{D_i}{k_B T}. \tag{2.13}$$

Thus, the mass and charge flux can be written as

$$\mathbf{N}_i = -D_i \nabla c_i - \frac{z_i e}{k_B T} D_i c_i \nabla \phi \tag{2.14}$$

$$\mathbf{J}_i = -z_i e D_i \nabla c_i - \sigma_i \nabla \phi. \tag{2.15}$$

The total current density in the electrolyte is obtained by the sum of the current fluxes of ions

$$\mathbf{i}_e = \sum_i \mathbf{J}_i = -\sum_i z_i e D_i \nabla c_i - \sigma \nabla \phi. \tag{2.16}$$

Defining the conductivity of the electrolyte as the sum of conductivities of the ions, $\sigma = \sum_i \sigma_i$. Electro neutrality is invoked to identify that $c_i = c_+ = c_- = c$. The current density is

$$\mathbf{i}_e = -e\nabla c \sum_i z_i D_i - \sigma \nabla \phi \tag{2.17}$$

$$= -\nabla c \sum_i z_i e D_i \left[1 + \frac{\partial \ln(\gamma_i)}{\partial \ln(c_i)}\right] - \sigma \nabla \phi \tag{2.18}$$

$$= -\nabla \ln(c) \frac{k_B T}{e} \sum_i \frac{\sigma_i}{z_i} \left[1 + \frac{\partial \ln(\gamma_i)}{\partial \ln(c_i)}\right] - \sigma \nabla \phi. \tag{2.19}$$

The final result is an expression for the total current density in the electrolyte in terms of the conductivity. Please note that the same expression, in terms of the diffusivities, is

$$\mathbf{i}_e = -e\nabla c \sum_i z_i D_i - \frac{e^2 c}{k_B T} \nabla \phi \sum_i z_i^2 D_i. \tag{2.20}$$

The first expression for the current flux is used to derive the current balance equation in the electrolyte, whereas the second is used to derive the mass conservation equation.

2.2.3 *Mass Conservation in the Electrolyte (Liquid Phase)*

To obtain the mass conservation equation, we start from

$$\mathbf{i}_e = -e\nabla c \sum_i z_i \tilde{D}_i - \frac{e^2 c}{k_B T} \nabla \phi \sum_i z_i^2 D_i. \tag{2.21}$$

From this equation, we can obtain expression for the electrolyte potential.

$$\nabla \phi = \frac{\mathbf{i}_e + e\nabla c \sum_i z_i \tilde{D}_i}{-\frac{e^2 c}{k_B T} \sum_i z_i^2 D_i} \tag{2.22}$$

The above expression is used to eliminate for the electrolyte potential from equation for the mass flux.

$$\mathbf{N}_i = -\tilde{D}_i \nabla c_i - \frac{z_i e}{k_B T} D_i c_i \left(\frac{\mathbf{i}_e + e \nabla c \sum_i z_i \tilde{D}_i}{-\frac{e^2 c}{k_B T} \sum_i z_i^2 D_i} \right) \tag{2.23}$$

which on simplification gives

$$\mathbf{N}_i = -\tilde{D}_i \nabla c_i + \frac{\frac{z_i e}{k_B T} D_i c}{\frac{e^2 c}{k_B T} \sum_i z_i^2 D_i} \left(\mathbf{i}_e + e \nabla c \sum_i z_i \tilde{D}_i \right) \tag{2.24}$$

The conservation equation for concentration states that the accumulation is accounted by the divergence of the flux and the rate of generation by charge transfer reaction.

$$\frac{\partial c_i}{\partial t} = -\nabla \cdot \mathbf{N}_i + R_i \tag{2.25}$$

For a porous electrode, the conservation equation is applied to the fractional volume occupied by the electrolyte. As the concentration is defined in terms of the pure electrolyte, the conservation equation is modified to be

$$\varepsilon \frac{\partial c_i}{\partial t} = -\nabla \cdot \mathbf{N}_i + R_i \tag{2.26}$$

The corresponding equation for the mass flux is given by

$$\mathbf{N}_i = -\varepsilon \mathscr{M}_i c_i \nabla \mu_i \tag{2.27}$$

This modification eventually results in

$$\mathbf{N}_i = -\varepsilon \tilde{D}_i \nabla c + \frac{\frac{z_i e}{k_B T} D_i c}{\frac{e^2 c}{k_B T} \sum_i z_i^2 D_i} \left(\mathbf{i}_e + e\varepsilon \nabla c \sum_i z_i \tilde{D}_i \right) \tag{2.28}$$

The mass flux has the corresponding divergence given by

$$\nabla \cdot \mathbf{N}_i = -\nabla \cdot \left(\varepsilon \tilde{D}_i \nabla c \right) + \nabla \cdot \left(\frac{z_i D_i}{e \sum_i z_i^2 D_i} \mathbf{i}_e \right) + \nabla \cdot \left(\frac{\varepsilon \nabla c z_i D_i \sum_i z_i \tilde{D}_i}{\sum_i z_i^2 D_i} \right) \tag{2.29}$$

This can be further simplified for univalent ions to give

$$\nabla \cdot \mathbf{N}_i = -\nabla \cdot \left[\varepsilon \nabla c \left(\tilde{D}_i - \frac{z_i D_i \sum_i z_i \tilde{D}_i}{\sum_i D_i} \right) \right] + \nabla \cdot \left(\frac{z_i D_i}{e \sum_i z_i^2 D_i} \mathbf{i}_e \right) \tag{2.30}$$

$$\nabla \cdot \mathbf{N}_i = -\nabla \cdot \left[\varepsilon \nabla c \left(\tilde{D}_i - \frac{z_i D_i \sum_i z_i \tilde{D}_i}{\sum_i D_i} \right) \right] + \nabla \cdot \left(\frac{z_i t_i}{e} \mathbf{i}_e \right) \tag{2.31}$$

As we are interested in the Li^+ ions, the equation results in

$$\nabla \cdot N_+ = -\nabla \cdot \left[\varepsilon \nabla c \left(t_- \tilde{D}_+ + t_+ \tilde{D}_- \right) \right] + \nabla \cdot \left(\frac{t_+}{e} \mathbf{i}_e \right) \tag{2.32}$$

Defining the total chemical diffusivity to be $\left(t_- \tilde{D}_+ + t_+ \tilde{D}_- \right) = \tilde{D}$, and t_-, t_+ are the transference numbers of the ions.

$$\nabla \cdot N_+ = -\nabla \cdot \left(\varepsilon \tilde{D} \nabla c \right) + \nabla \cdot \left(\frac{t_+}{e} \mathbf{i}_e \right) \tag{2.33}$$

Substituting above equation in the mass conservation equation,

$$\varepsilon \frac{\partial c}{\partial t} = \nabla \cdot \left(\varepsilon \tilde{D} \nabla c \right) - \nabla \cdot \left(\frac{t_+}{e} \mathbf{i}_e \right) - R_+ \tag{2.34}$$

The conservation equation can be written in molar units as

$$\varepsilon \frac{\partial c_e}{\partial t} = \nabla \cdot \left(\varepsilon D_e \nabla c_e \right) - \nabla \cdot \left(\frac{t_+}{F} \mathbf{i}_e \right) - R_+ \tag{2.35}$$

In order to avoid proliferation of variables $\mathbf{i}_e$ is considered to be in molar units, A/m^2. The rate of consumption is given by the Faradays law

$$R_+ = a_{p,n} j_{p,n} = \nabla \cdot \frac{\mathbf{i}_e}{F} \tag{2.36}$$

We will discuss the relevance of $a_{p,n}$ in a later section. The mass conservation equation is, finally

$$\varepsilon \frac{\partial c_e}{\partial t} = \nabla \cdot \left(\varepsilon D_e \nabla c_e \right) + \left(1 - t_+ \right) a_{p,n} j_{p,n} - \frac{\mathbf{i}_e}{F} \nabla \cdot t_+ \tag{2.37}$$

2.2.3.1 Application to Lithium-ion Cells

The analysis starts from the mass conservation, n for the electrolyte. The variable of interest is the lithium-ion concentration in the electrolyte, and c_e represents this variable. In any section of either of the electrodes, the local accumulation of lithium-ion concentration and transport due to diffusion is balanced by the rate of formation/dissipation of by the charge transfer reaction [32]. The mass balance is given by,

$$\varepsilon \frac{\partial c_e}{\partial t} = \nabla \cdot (\varepsilon D_e \nabla c_e) + (1 - t_+)\, a_{p,n} j_{p,n} - \frac{\mathbf{i}_e}{F} \nabla \cdot t_+ \tag{2.38}$$

In Eq. 2.38, D_e is the diffusivity of the electrolyte in the bulk solution, and can be a function of the electrolyte concentration. This equation accounts for the variation in the concentration lithium ion in the x-dimension. This dimension indicates the thickness of the cell, starts at the current collector of the negative electrode (anode during discharge) and ends at the positive electrode current collector. Thus, this model is a 1D model that accounts for concentration gradients across the cell thickness. The general mass conservation equation given by Eq. 2.38 needs to be modified to be of use in battery application.

In any electrode region, a fraction of volume filled by the electrolyte is given by the porosity, μ. To account for the composite nature of the electrode region, the effective diffusivity of the electrolyte is computed from the bulk value in the following manner:

$$D_{\text{eff}} = D_e \varepsilon^{\text{b}}. \tag{2.39}$$

Here, ε^{b} the Bruggmann factor, typically assigned the value of 1.5. Although the charge transfer reaction that occurs at the surface of the electrode and the electrolyte, and should enter the mass conservation equation as a boundary term. The porous electrode treats the electrode and electrolyte as continuum, this term enters the conservation equations as a source term. The mass flux of lithium due to this reaction is defined in terms of the surface area of the particles of the active material. To introduce this quantity into the conservation equation that is defined for a unit volume of the electrode, it is multiplied by the specific surface area per unit volume of the electrode (denoted by $a_{p,n}$), computed by assuming that all the particles are spheres of the same radius r_p,

$$a_{p,n} = \frac{4\pi r_p^2 N}{\frac{4}{3}\pi r_p^3 N}(1 - \varepsilon) = \frac{3(1-\varepsilon)}{r_p^2}. \tag{2.40}$$

This conservation equation discussed above applies to both the positive as well as the negative electrode regions individually. For either of the electrodes, the porosity, and hence the effective diffusivity, the transference number as well as the reaction rate are different. Conventionally, subscript p and n are used for the positive and negative electrode regions respectively, and the equations for each of these sections are

$$\varepsilon \frac{\partial c_e}{\partial t} = \nabla \cdot \left(\varepsilon D_{eff,p,n} \nabla c_e\right) + (1 - t_+)\, a_{p,n} j_{p,n} - \frac{\mathbf{i}_e}{F} \nabla \cdot t_+ \tag{2.41}$$

In the above equation, $D_{eff,p}$ and $D_{eff,n}$ are the effective diffusivities in the positive and negative electrode regions. As no reaction takes place in the separator region, the equation simplifies to

$$\varepsilon_s \frac{\partial c_e}{\partial t} = \nabla \cdot (D_s \nabla c_e)\,, \tag{2.42}$$

where the porosity of separator region corresponds to the partial volume occupied by the electrolyte within the pores of the separator matrix. It needs to be mentioned, however, that additional dependencies can be added to these base set of equations discussed in this section. For example, one can have a transference number or diffusivity that is dependent on concentration. Additionally, for high concentration solutions, the variation of these quantities can also be considered.

As lithium ions do not leave or enter the cell, the boundary condition for the lithium-ion mass conservation equation is zero mass flux at the current collector boundaries,

$$\left.\frac{\partial c_e}{\partial x}\right|_{x=0} = 0; \quad \left.\frac{\partial c_e}{\partial x}\right|_{x=L_{\text{cell}}} = 0, \tag{2.43}$$

where L_{cell} is the thickness of the cell. The continuity of concentrations and fluxes are also imposed at the positive electrode separator ($x = L_n$) as well as the separator-negative electrode interfaces ($x = L_n + L_s$).

2.2.4 *Charge Conservation in the Electrolyte (Liquid Phase)*

Corresponding to the mass flux, a current flux can be defined as

$$\mathbf{J}_i = z_i e \mathbf{N}_i. \tag{2.44}$$

The electric conductivity σ_i is defined as

$$\sigma_i = (z_i e)^2 \mathscr{M}_i c_i = (z_i e)^2 c_i \frac{D_i}{k_B T}. \tag{2.45}$$

Thus, charge flux can be written as

$$\mathbf{J}_i = -z_i e D_i \nabla c_i - \sigma_i \nabla \phi. \tag{2.46}$$

The total current density in the electrolyte is obtained by the sum of the current fluxes of ions

$$\mathbf{i}_e = \sum_i \mathbf{J}_i = -\sum_i z_i e D_i \nabla c_i - \sigma \nabla \phi. \tag{2.47}$$

Defining the conductivity of the electrolyte as the sum of conductivities of the ions, $\sigma = \sum_i \sigma_i$. Electro neutrality is invoked to identify that $c_i = c_+ = c_- = c$. The current density is

$$\mathbf{i}_e = -e\nabla c \sum_i z_i D_i - \sigma \nabla \phi \tag{2.48}$$

$$= -\nabla c \sum_i z_i e D_i \left[1 + \frac{\partial \ln(\gamma_i)}{\partial \ln(c_i)}\right] - \sigma \nabla \phi \tag{2.49}$$

$$= -\nabla \ln(c) \frac{k_B T}{e} \sum_i \frac{\sigma_i}{z_i} \left[1 + \frac{\partial \ln(\gamma_i)}{\partial \ln(c_i)}\right] - \sigma \nabla \phi. \tag{2.50}$$

The final result is an expression for the total current density in the electrolyte in terms of the conductivity. Now

$$\mathbf{i}_e = -\nabla \ln(c) \frac{k_B T}{e} \sum_i \frac{\sigma_i}{z_i} \left[1 + \frac{\partial \ln(\gamma_i)}{\partial \ln(c_i)}\right] - \sigma \nabla \phi. \tag{2.51}$$

The transference number is defined in terms of conductivities as $t_i = \sigma_i/\sigma$ and satisfies the relation $t_+ + t_- = 1$. The current density can be rewritten as

$$\mathbf{i}_e = -\nabla \ln(c) \frac{k_B T}{e} \sum_i \frac{t_i}{z_i} \left[1 + \frac{\partial \ln(\gamma_i)}{\partial \ln(c_i)}\right] - \sigma \nabla \phi. \tag{2.52}$$

For a binary electrolyte, like in the case of lithium-ion cells, $z_+ = z_- = 1$.

$$\mathbf{i}_e = \nabla \ln(c) \frac{k_B T \sigma}{e} \left((1 - t_+) \left[1 + \frac{\partial \ln(\gamma_-)}{\partial \ln(c_i)}\right] - t_+ \left[1 + \frac{\partial \ln(\gamma_-)}{\partial \ln(c_i)}\right]\right) - \sigma \nabla \phi. \tag{2.53}$$

On simplification

$$\mathbf{i}_e = \nabla \ln(c) \frac{k_B T \sigma}{e} \left(1 + \frac{\partial \ln(\gamma_-)}{\partial \ln(c)} - 2t_+ - t_+ \left[\frac{\partial \ln(\gamma_+)}{\partial \ln(c)} + \frac{\partial \ln(\gamma_-)}{\partial \ln(c)}\right]\right) - \sigma \nabla \phi. \tag{2.54}$$

The activity coefficient of the electrolyte can defined in terms of the respective ionic values using $\gamma_\pm = \sqrt{\gamma_+ \gamma_-}$. This leads to further simplification as given below

$$\mathbf{i}_e = \nabla \ln(c) \frac{k_B T \sigma}{e} \left(2(1 - t_+) \left[1 + \frac{\partial \ln(\gamma_\pm)}{\partial \ln(c)}\right] - \left[1 + \frac{\partial \ln(\gamma_+)}{\partial \ln(c)}\right]\right) - \sigma \nabla \phi. \tag{2.55}$$

This results in the final expression for the current density in the electrolyte

$$\mathbf{i}_e = -\sigma \nabla \phi + 2(1 - t_+) \frac{k_B T \sigma}{e} \left[1 + \frac{\partial \ln(\gamma_\pm)}{\partial \ln(c)}\right] \nabla \ln(c) - \frac{k_B T \sigma}{e} \left[1 + \frac{\partial \ln(\gamma_+)}{\partial \ln(c)}\right] \nabla \ln(c). \tag{2.56}$$

The last term can be written in terms of the mass diffusivity as

$$\mathbf{i}_e = -\sigma\nabla\phi + 2(1-t_+)\frac{k_B T\sigma}{e}\left[1+\frac{\partial\ln(\gamma_\pm)}{\partial\ln(c)}\right]\nabla\ln(c) - e\nabla c\frac{D_+}{t_+}\left[1+\frac{\partial\ln(\gamma_+)}{\partial\ln(c)}\right]. \quad (2.57)$$

For binary electrolytes with unit charges, the transference number can also be defined as $t_i = D_i/\sum D_i = D_i/D$. Hence

$$\mathbf{i}_e = -\sigma\nabla\phi + 2(1-t_+)\frac{k_B T\sigma}{e}\left[1+\frac{\partial\ln(\gamma_\pm)}{\partial\ln(c)}\right]\nabla\ln(c) - e\nabla cD\left[1+\frac{\partial\ln(\gamma_+)}{\partial\ln(c)}\right]. \quad (2.58)$$

Defining a chemical potential to be $\tilde{\mu}_i = k_B T\ln(\gamma_i c_i)$, following simplification,

$$\mathbf{i}_e = -\sigma\nabla\phi + 2(1-t_+)\frac{k_B T\sigma}{e}\left[1+\frac{\partial\ln(\gamma_\pm)}{\partial\ln(c)}\right]\nabla\ln(c) - \frac{ec}{k_B T}D\nabla\tilde{\mu}_i. \quad (2.59)$$

In obtaining the electrolyte current balance [32], one needs to assume that the flux generated due to coupling between the total mass diffusivity of the electrolyte and the chemical potential gradients are neglected, in comparison to the other contributions. Thus, the equation simplifies to

$$\mathbf{i}_e = -\sigma\nabla\phi + 2(1-t_+)\frac{k_B T\sigma}{e}\left[1+\frac{\partial\ln(\gamma_\pm)}{\partial\ln(c)}\right]\nabla\ln(c). \quad (2.60)$$

The above equation is written for the electrolyte , either in the electrode region or the separator. This equation is further refined on identifying $e/(k_B T) = F/(R_G T)$,

$$\mathbf{i}_e = -\sigma\nabla\phi + 2(1-t_+)\frac{R_G T\sigma}{F}\left[1+\frac{\partial\ln(\gamma_\pm)}{\partial\ln(c)}\right]\nabla\ln(c). \quad (2.61)$$

2.2.4.1 Application to Lithium-Ion Cells

The electrolyte current is computed using the concentrated solution theory as discussed above. Conventionally, Eq. 2.61 is written as

$$i_e = -\kappa_i\nabla\phi_e + \kappa_i\frac{2R_G T}{F}(1-t_i)\left(1+\frac{\ln f_i}{\ln c_e}\right)\nabla\ln c_e, \quad (2.62)$$

where κ_i is the effective conductivity of the electrolyte computed from the porosities of either the positive or the negative electrode region, using the Bruggmann relation (Eq. 2.39). This equation is applied for each of the electrodes to solve for the electrolyte potential, ϕ_e. As cell potential is arbitrary up to a constant, a boundary condition is to ground the negative end of the cell ($x = 0$), i.e., $\phi_e = 0$. As the current enters and leaves the cell through the solid particles in contact with the current collectors, an insulation boundary condition is set at the other end of the cell,

$\left.\frac{\partial \phi_e}{\partial x}\right|_{x=L_{\text{cell}}}$. Alternate is to set insulation boundary condition at both $x = 0$ and $x = L_{\text{cell}}$.

Since all the current flows right through the separator region, the equation becomes

$$I = -\kappa_s \nabla \phi_e + \kappa_s \frac{2R_G T}{F} (1 - t_s) \left(1 + \frac{\ln f_s}{\ln c_e}\right) \nabla \ln c_e, \tag{2.63}$$

where I is the total current that enters or leaves the cell.

2.2.5 *Charge Conservation in the Electrode (Solid) Region*

The conservation of current in the solid phase of the electrode region is provided by the generalized Ohm's law written as

$$\mathbf{i}_s = -\sigma_s \nabla \phi_s. \tag{2.64}$$

Throughout the negative electrode lithium ions are generated due to the charge transfer reaction and are consumed in the positive electrode. Faradays law gives the equivalent current generated due the production of lithium ions. For one mole of univalent lithium ion, the following relation between the divergence of the solid phase current and the rate of generation of lithium ions holds

$$\nabla \cdot \mathbf{i}_s = F a_i j_i. \tag{2.65}$$

Combining the Faradays law with the Ohms law relates the potential in the electrode phase with the rate of reaction.

$$-\sigma_i \nabla^2 \phi_s = F a_i j_i. \tag{2.66}$$

The above equation is solved for the positive and negative electrodes to obtain the potential of the solid phase in the electrode. As the current enters the cell at $x = 0$ and leaves at $x = L_{\text{cell}}$, boundary conditions are

$$-\sigma_n \nabla \phi_s|_{x=0} = I; \quad -\sigma_n \nabla \phi_s|_{x=L_{\text{cell}}} = I. \tag{2.67}$$

An alternate boundary condition for the solid potential is

$$\phi_s|_{x=0} = 0. \tag{2.68}$$

This boundary condition can be used to set the datum in the potential, if insulation boundary conditions at used for the electrolyte potential, ϕ_l are $x = 0$ and $x = L_{\text{cell}}$. At $x = L_n$, the current leaves the solid phase and is the total applied current is carried through the electrolyte in the separator, and the reverse happens at $x = L_n + L_s$. Thus, the solid phase current is zero at these interfaces

$$-\sigma_n \nabla \phi_s|_{x=L_n} = 0; \quad -\sigma_p \nabla \phi_s|_{x=L_n+L_s} = 0. \tag{2.69}$$

The solid phase current conservation equation is solved for both the electrodes to obtain the potential in the solid phase at both the electrodes. The solution also enables to study the variation of this quantity at various cell thickness. Solution of this equation is of special significance, as the cell voltage, the final output of any mathematical model for the battery is defined as the difference in solid phase potential between the ends of the cell.

$$V_{\text{cell}} = \phi_s|_{x=L_{\text{cell}}} - \phi_s|_{x=0}. \tag{2.70}$$

2.3 The Charge Transfer Reaction

The charge transfer reaction involves an equilibrium component as well as a dynamic component. The equilibrium component based on thermodynamics relates the open-circuit potential to the activity or equivalently the concentrations, and is given by the Nernst equation.The dynamic component relates the driving force for the charge transfer reaction to the reaction flux or the current transferred. We will start with the discussion of the Nernst equation.

2.3.1 Nernst Equation: Equilibrium Component and its Thermodynamic Connect

For any charge transfer reaction at equilibrium,

$$R + ne \rightleftharpoons P \tag{2.71}$$

The Nernst equation relates the equilibrium electrode potential V_0 to the standard cell potential $V_0^{'}$ and the compositions of the reactants and the products.

$$V_0 = V_0^{'} + \frac{R_G T}{nF} \ln Z \tag{2.72}$$

The compositions are adequately represented Z, the ratio of the activities of the products to the reactants, or simply, the ratio of their concentrations,

$$Z = \frac{a_p}{a_R} = \frac{c_P}{c_R}. \tag{2.73}$$

To obtain the Nernst equation, the first step is to identify the fundamental connection between electrochemical and thermodynamic quantities. Once this is identified, fundamental thermodynamic relations are subsequently used. The most intriguing thermodynamic quantity is the Gibbs free energy, G. This quantity being defined under conditions of constant temperature and pressure, easily achievable experimental conditions, makes it the most used (and misunderstood!) thermodynamic variable. The Gibbs free energy is defined as

$$G = H + TS. \tag{2.74}$$

We will go through a simple thermodynamic relation and see what measure G corresponds to. The enthalpy H is related to internal energy E, and pressure P and volume V, via

$$H = E + PV. \tag{2.75}$$

The change in enthalpy for a constant pressure process is given by,

$$\Delta H = \Delta E + \Delta(PV) = \Delta E + P\Delta V. \tag{2.76}$$

The change in internal energy can also be related to heat transfer and the reversible work,

$$\Delta E = \Delta Q - \Delta W = T\Delta S - \Delta W_{\text{rev}}, \tag{2.77}$$

with S being the entropy.

From the earlier relations, the change in Gibbs free energy at constant temperature is

$$\Delta G = \Delta H - \Delta(TS) = P\Delta V - \Delta W_{\text{rev}}. \tag{2.78}$$

Total reversible work consists of expansion work given by $P\Delta V$ and all other types of work. Thus, the difference between the quantities in the right side of the above equations gives the net non-expansion work that is derived out of the system.

In the case of the lithium-ion cell it is electrical work, ΔW_{ec}. Thus, it can be seen that

$$\Delta G = -\Delta W_{\text{ec}}. \tag{2.79}$$

Change in Gibbs free energy gives the reversible electrical work in a lithium-ion cell. Electrical work is also defined in terms of transferring nF electrons across an equilibrium voltage of V_0 as,

$$\Delta W_{\text{ec}} = nFV_0. \tag{2.80}$$

Thus, we arrive at the fundamental relationship between thermodynamics and electrochemistry that relates the equilibrium potential and Gibbs free energy,

$$\Delta G = -nFV_0. \tag{2.81}$$

For chemical reaction thermodynamics, change in Gibbs free energy is related to concentrations as,

$$\Delta G = \Delta G(0) + R_G T \ln \frac{c_P}{c_R}. \tag{2.82}$$

which again can be derived from identifying that ΔG is the difference of chemical potentials of the product and the reactant.

$$\Delta G = \mu_P - \mu_R. \tag{2.83}$$

As seen earlier, chemical potential is defined for moderate concentrations as

$$\begin{aligned} \mu_P &= \mu_P^{'} + R_G T \ln c_P, \\ \mu_R &= \mu_R^{'} + R_G T \ln c_R. \end{aligned} \tag{2.84}$$

Substituting this in the earlier equation, one gets the relation for change in Gibbs free energy in terms of concentrations.

Using our newfound relation between Gibbs free energy and equilibrium potential and dividing the above equation by nF, we get the Nernst equation,

$$V_0 = V_0(0) + \frac{R_G T}{nF} \ln \frac{c_P}{c_R}. \tag{2.85}$$

The Nernst equation is the basis for obtaining any dependence of equilibrium potential of positive or negative electrode with the concentrations. This equation, however, obtained from thermodynamics, gives only an ideal case. However, material variations impart unique features in the OCV curve. For realistic scenarios, various modifications are proposed. In Figs. 2.3 and 2.4, the open-circuit potentials of common materials are shown. The present crop of commercial batteries predominantly use some blend of carbon (graphite, coke, hard carbon, or mesocarbon microbeads (MCMB)). There are more variety for the cathode electrodes and it is dependent on the application and the requisite energy density.With this subsection, we obtain the equilibrium part in the charge transfer reaction. The dynamic part is subsequently derived.

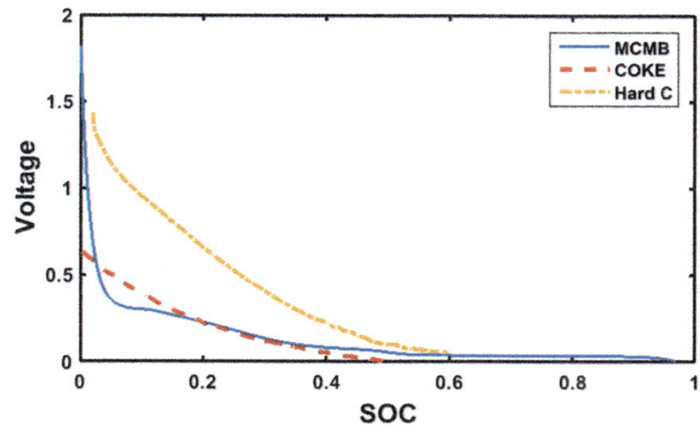

Fig. 2.3 The open-circuit potential of commonly used anode materials

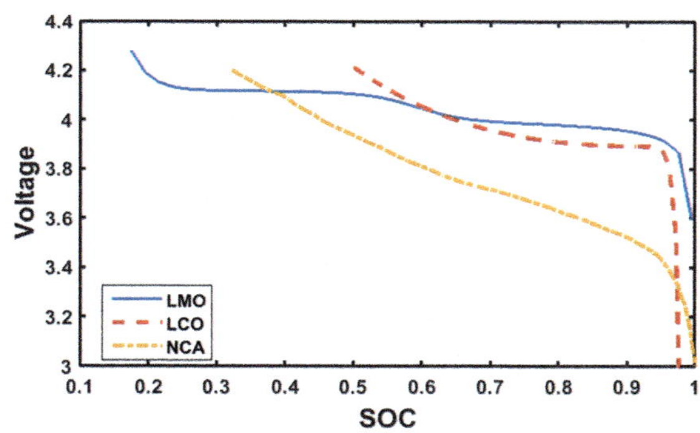

Fig. 2.4 The open-circuit potential of commonly used cathode materials

2.3.2 *Butler–Volmer Equation: The Dynamic Component*

For a charge transfer chemical reaction as shown below, R is reduced and P is oxidized. The former is termed as an anodic process, and the latter cathodic. The chemical reaction results in an equivalent current density given by,

$$\begin{aligned} j_c &= k_c c_R \quad \text{and,} \\ j_a &= k_a c_P. \end{aligned} \tag{2.86}$$

where j_c and j_a are termed as the cathodic and anodic current densities, k_c and k_a are the rate constants for the respective reactions. These reactions happen at the electrode surfaces, and the net reaction rate determines if the reversible process is oxidation or reduction. In the negative electrode, the net reaction during discharge is oxidation and the net reaction at the positive electrode is reduction. During charging however, the processes are reversed. As a result reduction reaction occurrs at the negative electrode and oxidation at the positive electrode. The net reaction rate that results in the net current density is given by

$$i = Fj_a - Fj_c = Fk_a c_P - Fk_c c_R \tag{2.87}$$

Each of these processes is activated process. This means that the processes are triggered once the molecules undergo a threshold internal energy change required at a given temperature. Thus

$$k = k_0^{'} \exp -\Delta E/R_{\mathrm{G}}T = k_0^{'} \exp\{-\Delta H/R_{\mathrm{G}}T\} \tag{2.88}$$

Because, from first law of thermodynamics, $\Delta H \equiv \Delta E$, with negligible change in P or V. This equation is modified and using the second law, Gibbs free energy again appears

$$k = k_0^{'} \exp\{-(\Delta H - T\Delta S)/R_{\mathrm{G}}T\} = k_0^{'} \exp\{-\Delta G/R_{\mathrm{G}}T\}. \tag{2.89}$$

Thus, the rate constants of both the electrodic process can be expressed in terms of Gibbs free energy as,

$$\begin{aligned} k_c &= k_{c0}^{'} \exp\{-(\Delta G_c)/R_{\mathrm{G}}T\}, \\ k_a &= k_{a0}^{'} \exp\{-\Delta G_a/R_{\mathrm{G}}T\}. \end{aligned} \tag{2.90}$$

It can be inferred from our earlier discussion earlier, Gibbs free energy pertains to the electrical work needed to switch on these processes. The net current density at any electrode is given by

$$i = Fk_{a0}^{'} \exp\{-\Delta G_a/R_{\mathrm{G}}T\}\, c_P - Fk_{c0}^{'} \exp\{-(\Delta G_c)/R_{\mathrm{G}}T\}\, c_R. \tag{2.91}$$

The above equation, the net rate of a reversible reaction that is activated, is the basic form of Butler–Volmer reaction. The reactions are however, not at equilibrium, which was our basic premise for moving beyond Nernst type equation.

In lithium-ion batteries, a lithium-ion takes up only one electron during charge transfer reaction,the reaction is like, $R + e \rightleftharpoons P$. If the reaction involves a change of potential given by $\Delta\Phi = V - V(0)$, the extra electrical work required, the Gibbs free energy change is given by

$$\Delta G = \Delta G(0) + F\Delta\Phi. \tag{2.92}$$

If the energy state at the initial state (denoted as '0') is itself enough for the reaction to occur, however, $\Delta G = \Delta G(0)$. In a realistic scenario, the energy change required is taken as an average of these two extremes, and the constant, a transfer coefficient α is introduced,

$$\Delta G = \Delta G(0) + \alpha F\Delta\Phi. \tag{2.93}$$

The anodic and cathodic processes are the forward and reverse reactions of the reversible electrode reaction. If we consider the above scenario for the cathodic process, corresponding equation for the anodic process gives

$$\Delta G = \Delta G(0) - (1-\alpha)F\Delta\Phi. \tag{2.94}$$

Inserting the above expressions for our net rate of reaction, and absorbing the terms involving $\Delta G(0)$ into the constant,

$$i = Fk_{a0} \exp\{(1-\alpha)F\Delta\Phi/R_{\mathrm{G}}T\}\, c_P - Fk_{c0} \exp\{-\alpha F\Delta\Phi/R_{\mathrm{G}}T\}\, c_R. \tag{2.95}$$

For a system at equilibrium, $\Delta\Phi = V_0 - V(0)$, and the net current is zero. This results in both the current densities being the same, termed as the exchange current density, i_0. Let us also introduce $f = F/R_{\mathrm{G}}T$

$$i_0 = Fk_{a0} \exp\{(1-\alpha)F\Delta\Phi/R_{\mathrm{G}}T\}\, c_{P,EQ} = Fk_{c0} \exp\{-\alpha F\Delta\Phi/R_{\mathrm{G}}T\}\, c_{R,EQ}. \tag{2.96}$$

At this state of equilibrium, one can use Nernst equation to see the dependence between concentrations,

$$\begin{aligned} \exp\{f(V_0 - V(0))\} &= \frac{c_{P,Eq}}{c_{R,Eq}}, \\ \exp\{-\alpha f(V_0 - V(0))\} &= \left(\frac{c_{P,Eq}}{c_{R,Eq}}\right)^{-\alpha}. \end{aligned} \tag{2.97}$$

Thus, the exchange current density is written as

$$i_0 = Fk_0 c_{P,Eq}^{\alpha} c_{R,Eq}^{(1-\alpha)} \tag{2.98}$$

To eliminate the effect of the datum potential, the net rate is divided by the exchange current,

$$\frac{i}{i_0} = \left[\frac{Fk_{a0}c_P \exp\{(1-\alpha)f(V_0 - V(0))\}}{Fk_{a0}c_{P,Eq} \exp\{(1-\alpha)f(V_0 - V(0))\}} - \frac{Fk_{c0}c_R \exp\{-\alpha f(V_0 - V(0))\}}{Fk_{c0}c_{R,Eq} \exp\{-\alpha f(V_0 - V(0))\}}\right] \tag{2.99}$$

Eventually, the most common form of the Butler–Volmer reaction flux is obtained as

$$i = i_0 \left[\frac{c_P}{c_{P,Eq}} \exp\{(1-\alpha)f(V_0 - V(0))\} - \frac{c_R}{c_{R,Eq}} \exp\{-\alpha f(V_0 - V(0))\}\right]. \tag{2.100}$$

In many cases, a local equilibrium of concentration is considered, which further simplifies the equation to give,

$$i = i_0 \left[\exp\{(1-\alpha)f(V_0 - V(0))\} - \exp\{-\alpha f(V_0 - V(0))\}\right] \tag{2.101}$$

2.3.2.1 Application to Lithium-Ion Cells

Lithium ions in the electrolyte with concentration c_e, and potential ϕ_e intercalate into the electrode particles. In this case, the concentration of the free active sites also determines the rate of reaction. In an electrode particle with solid concentration c_s

and the solubility $c_{s,max}$, the concentration of unoccupied sites is given by $c_{s,max} - c_s$. Also, due to the electrolyte potential, $\Delta\phi = \phi_s - \phi_e - V(0)$. Accounting for these factors, the final form of Butler–Volmer equation that is used to solve the electrochemical model reads as,

$$i = i_0 \left[\exp\left\{(1-\alpha) f\left(\phi_s - \phi_e - V(0)\right)\right\} - \exp\left\{-\alpha f\left(\phi_s - \phi_e - V(0)\right)\right\}\right], \quad (2.102)$$

With the exchange current density given by,

$$i_0 = F k_0 c_s^{\alpha} \left(c_{s,max} - c_s\right)^{1-\alpha} c_e^{(1-\alpha)}. \quad (2.103)$$

To incorporate in the electrochemical model, this equation has to be solved for both the electrodes. While solving for realistic case scenarios, the rate constants of the electrodes are often adjusted to match the experimental results.

Chapter 3
Theoretical Framework of Electrochemical–Thermal Model (ECT)

The energy balance for a Li-ion cell is derived from principles of nonequilibrium thermodynamics. The equations are integrated into the electrochemical model. The model is extended to special cases like electrodes with phase change. Performance indicators that can be computed from the model that enables an optimal design are developed.

3.1 Energy Balance for Lithium Ion Cells: Lumped Heat Source

In this section, we will derive the energy balance for a lithium ion cell. The principle cause of heat release is the charge transfer reaction. Before we venture into this, we need to get some dependencies that thermodynamics can help us obtain. Specifically, we are interested in the Gibbs free energy. We saw in the last section that change in the Gibbs free energy gives the maximum non expansion work that can be obtained from the system. Lithium ion cell is an electrochemical system with no expansion work and thus change in the Gibbs free energy is the central to the various inter relations that can be drawn. The heat released during a chemical reaction is related to the change in this quantity through the Gibbs Helmholtz equation given by:

$$\Delta H = -T^2 \frac{d\Delta G}{dT} \tag{3.1}$$

As we saw earlier, $\Delta G = -nFV_0$ and hence the heat of reaction is related to the OCV:

$$\Delta H = nFT^2 \frac{dV_0}{dT} \tag{3.2}$$

The heat of reaction per unit mole (or equivalently Coulomb) is given by

$$\frac{\Delta H}{nF} = -V_0 + T\frac{dV_0}{dT} \tag{3.3}$$

K. S. Hariharan et al., *Mathematical Modeling of Lithium Batteries*,
Green Energy and Technology, https://doi.org/10.1007/978-3-319-03527-7_3

The charge transfer reaction generates lithium ions equivalent to a current flux of I_{BV} and the heat generated due to the charge transfer reaction is given by:

$$Q_{BV} = I_{BV} \frac{\Delta H}{nF} \tag{3.4}$$

As seen earlier, the relation between the OCV and the heat of reaction results in the following equation:

$$Q_{BV} = I_{BV} \left(-V_0 + T \frac{dV_0}{dT} \right) \tag{3.5}$$

This equation has two terms, and correspond to two quantities. The first term is the work that one can get out of the electrochemical cell. This work being reversible, is also the maximum work the cell generates, and can be written in terms of the measurable cell current and the OCV.

$$I_{BV} V_0 = I V_0 \tag{3.6}$$

The second term, $I_{BV} T \frac{dV_0}{dT}$ is related to the entropic contribution. To understand this, we would need to visit basic thermodynamics again, from change in Gibbs free energy is:

$$\Delta G = \Delta H - \Delta(ST). \tag{3.7}$$

For a reversible process at constant pressure, $\Delta H = \Delta Q_{rev} = T \Delta S$, thus $\Delta G = -S \Delta T$.

This results in revealing the relation between the temperature dependence of Gibbs free energy and change in entropy

$$\frac{\partial \Delta G}{\partial T} = -\Delta S \tag{3.8}$$

As we have seen that the $\Delta G = -nFV_0$, and hence the temperature derivative of the OCV is given by the entropy per mole (or Coulomb)

$$\frac{\partial V_0}{\partial T} = \frac{\Delta S}{nF} \tag{3.9}$$

It follows that the second term in the heat release term is the entropy change due to the charge transfer reaction. This analysis tells us that the charge transfer reaction generates heat that can be separated to maximum equilibrium work component as well as the entropic contributions.

As the total heat generated by the reaction is known, one can write an energy conservation equation for the lithium ion cell. A lithium ion cell at current I and voltage V generates an irreversible work component I V. From the heat generated due to the reaction, as seen above, generates a reversible work component as well as an entropic component. Additionally, the heat is taken from the system to ambient, typically by convection. The energy conservation is written as:

$$\rho C_P \frac{\partial T}{\partial t} = IV + Q_{BV} + Q_{Conv} \tag{3.10}$$

The heat loss by convection is written in terms of the heat transfer coefficient h and area A, and ambient temperature as given by

$$Q_{Conv} = -hA(T - T_{Amb}) \tag{3.11}$$

The complete energy conservation now becomes:

$$\rho C_P \frac{\partial T}{\partial t} = [IV - I_{BV} V_0] + I_{BV} T \frac{dV_0}{dT} - hA\,(T - T_{Amb}) \tag{3.12}$$

In a lumped model, the current generated by the reaction is the cell current, and hence:

$$\rho C_P \frac{\partial T}{\partial t} = I\,[I - V_0] + IT \frac{dV_0}{dT} - hA\,(T - T_{Amb}) \tag{3.13}$$

The first term on the right side, the difference between the irreversible work and the reversible work quantifies total irreversible loss during the cell operation. The irreversible losses are due to the transport and conduction processes, and results in exothermic heat generation during charge or discharge. The second term corresponds to the reversible losses, and can be recovered in a complete charge–discharge cycle. The energy conservation takes the form:

$$\rho C_P \frac{\partial T}{\partial t} = Q_{irr} + Q_{rev} + Q_{Conv} \tag{3.14}$$

3.2 Energy Balance for Lithium Ion Cells: Distributed Heat Source

The energy conservation discussed in the earlier section can be used to estimate the temperature rise of the cell at various stages of cell operation. This approach however, cannot be used to distinguish the sources of heat generation between the electrodes. For efficient cell as well as cooling system design, it is important to identify the prominent sources of heat at the operating conditions that result in temperature release.

To enquire into these details, one needs to develop the energy conservation for lithium ion batteries based on a distributed heat source approach. In this approach, one looks at the interface of the solid electrode particle and the electrolyte. It is at this interface that the charge transfer reaction occurs, producing heat. The overpotential being the driving force for the reaction, and the local current density the effect, the heat released is given by the product of these two terms. Additionally, the current flux at the interface creates additional generation of heat, the effect called as Peltier effect. Thus the total heat generated due to the charge transfer reaction is given by:

$$q_{BV} = Fj_{p,n}\left[\eta_{p,n} + \Pi_{p,n}\right] \tag{3.15}$$

where the overpotential is given by

$$\eta_{p,n} = (\phi_s - \phi_e - V_0) \tag{3.16}$$

Please note that this heat is generated in both the electrodes, and are denoted by the subscripts p or n respectively. The first term, being always an exothermic terms results in the irreversible contribution, and the second that involves the peltier effect is the reversible contribution, and will depend on whether the cell is charging or discharging. The Peltier effect needs a bit of understanding, and is addressed by basic theorems on nonequilibrium thermodynamics.

3.3 Basics of Nonequilibrium Thermodynamics

Details of nonequilibrium thermodynamics are clearly out of scope of this book. An attempt however, is made to such an extent that the Peltier effect be understood, based on the Onsager-DeGroot-Callen method. Classical thermodynamic relations are extended to irreversible scenarios that relate the various fluxes. Next, flux force relations are drawn, based on the equilibrium principles to obtain the exact coupling coefficients. Onsager's principle states that any flux can be coupled with any forces, and we deal with the fluxes of particles J_N and energy J_E. This is because the potentials and the forces are established for these fluxes. The corresponding potentials are given by $-\mu/T$ and 1/T respectively, and hence the forces by $\nabla(-\mu/T)$ and $\nabla(1/T)$. The particle and energy fluxes are given by:

$$J_N = L_{NN}\nabla(-\mu/T) + L_{NE}\nabla(1/T) \tag{3.17}$$

$$J_E = L_{EN}\nabla(-\mu/T) + L_{EE}\nabla(1/T) \tag{3.18}$$

Please note that μ is the electrochemical potential of the system. The motive is to obtain an expression for the Peltier coefficient that couples the electric and heat fluxes, and hence we need to get the corresponding fluxes. The current flux is given directly from the particle flux, number of moles and the Faradays constant as $J = nFJ_N$. The heat flux, J_Q is obtained, based on the first laws of thermodynamics to be:

$$J_Q = J_E - \mu J_N \tag{3.19}$$

From the earlier relations, this can be written in terms of the driving forces, and after some simplification results in:

$$J_Q = -\frac{1}{T}\nabla\mu\,[L_{NE} - \mu L_{NN}] + \nabla(1/T)\left[L_{EE} - \mu(L_{NE} + L_{EN}) + \mu^2 L_{NN}\right] \tag{3.20}$$

The equations for heat and current flux results in the Peltier coefficient as:

$$\Pi = \left(\frac{J_Q}{J}\right)_{\nabla T=0} = \frac{[L_{NE} - \mu L_{NN}]}{nFL_{NN}} \tag{3.21}$$

At this stage, the coupling coefficients are not known, and the form is not of great usage. From the definition of entropy in the equilibrium framework, $dS = dQ_{rev}/T$ analogous form for the entropy flux can be obtained in terms of the heat flux as:

$$J_S = J_Q/T \tag{3.22}$$

From the earlier obtained equation for the heat flux, the entropy flux becomes:

$$J_S = -\frac{1}{T^2}\nabla\mu\,[L_{NE} - \mu L_{NN}] + \frac{1}{T}\nabla(1/T)\left[L_{EE} - \mu(L_{NE} + L_{EN}) + \mu^2 L_{NN}\right] \tag{3.23}$$

To see this equation in the right perspective we notice that, the current flux without any temperature gradients, is given by:

$$J_{\nabla T=0} = -\frac{nFL_{NN}}{T}\nabla\mu \tag{3.24}$$

This equation, the Ohms law, can be inverted to express the electrochemical potential in terms of the current flux density:

$$\nabla\mu = -\frac{TJ_{\nabla T=0}}{nFL_{NN}} \tag{3.25}$$

Resulting in replacement of this quantity in the entropy flux that now reads:

$$J_S = \frac{[L_{NE} - \mu L_{NN}]}{nFTL_{NN}} J_{\nabla T=0} + \frac{1}{T}\left[L_{EE} - \mu(L_{NE} + L_{EN})\mu^2 L_{NN}\right]\nabla(1/T) \tag{3.26}$$

The above equations shows that the entropy flux has two parts. In the first term, entropy is carried by particle flux, driven by electrochemical driving forces. The second contribution is due to thermal gradients. From the construct of this equation, it can be inferred that the entropy change per mole of charged species is given by:

$$\Delta S_N = \frac{[L_{NE} - \mu L_{NN}]}{nFTL_{NN}} \tag{3.27}$$

Comparing the above expression for entropy and the one derived earlier for the Peltier coefficient, it can be seen that

$$\Pi = \frac{T\Delta S_N}{nF} \tag{3.28}$$

We had seen earlier that the change in entropy is related to the temperature dependence of the Gibbs energy which in turn to the OCV. Thus,

$$\Pi = \frac{T\Delta S_N}{nF} = -\frac{T}{nF}\frac{\partial \Delta G}{\partial T} = T\frac{\partial V_0}{\partial T} \tag{3.29}$$

With this insight, the heat released at the interface can be written as:

$$q_{BV} = Fj_{p,n}\left[\eta_{p,n} + T\frac{\partial V_{0,p,n}}{\partial T}\right] \tag{3.30}$$

with

$$\Pi_{p,n} = \frac{T\Delta S_{p,n}}{nF}; \quad \Delta S_{p,n} = nF\frac{\partial V_{0,p,n}}{\partial T} \tag{3.31}$$

With the principal heat sources known, we return to the energy conservation equation.

3.4 The Energy Conservation Equation with Distributed Heat Sources

In the macro-homogeneous framework, every point across the electrode thickness is considered to be juxtaposition of the electrode and the electrolyte phases, and they are not distinguished. Due to this reason, the heat flux terms that we discussed can be introduced in the energy conservation equation by multiplying with the surface to volume ratio of the electrode. It is to be noted that in the energy conservation equation, the hear generation per unit volume is considered.

In addition to the principal heat sources due to the Butler-Volmer reaction, in some cells, the electrode can undergo phase transitions resulting in additional sources of heat,

$$q_{\alpha\to\beta} = \Delta h\frac{\partial\left(\frac{r_i}{R}\right)}{\partial t} \tag{3.32}$$

Where Δh is the difference in the difference in the enthalpy between the two phases given by

$$\Delta h = \left(h_\alpha - h_\beta\right) + \left(C_{p,\alpha} - C_{p,\beta}\right)T \tag{3.33}$$

Combining all these we can write a general energy conservation equation per unit volume of the electrode as:

$$\rho C_P \frac{\partial T}{\partial t} = -\Delta \cdot q + H_S \Delta \cdot N_S + H_e \Delta \cdot N_e + a_S q_{BV} + q_{\alpha \to \beta} \tag{3.34}$$

Of which we are now familiar with the last two heat generation terms, but other quantities, that pertain to the manner in which secondary heat is generated due to current conduction, need more attention. Heat is carried in the cell due to conduction, and with the mass flux in the electrode and the electrolyte regions.

$$q = -\kappa \nabla T + H_S N_S + H_e N_e \tag{3.35}$$

By substitution in the energy balance,

$$\rho C_P \frac{\partial T}{\partial t} = \kappa \nabla^2 T - N_S \cdot \nabla H_S - N_e \cdot \nabla H_e + a_S q_{BV} + q_{\alpha \to \beta} \tag{3.36}$$

By applying Gibbs-Helmoltz equation per unit mole basis,

$$H = -T^2 \frac{\partial}{\partial T} \left(\frac{\mu}{T} \right) = \mu - T \frac{\partial \mu}{\partial T} \tag{3.37}$$

The electrochemical potential can be expressed in terms of the activity coefficient, and potential as:

$$\mu = \mu^{'} R_G T \ln \gamma + nF\phi \tag{3.38}$$

Ignoring the temperature dependence of activity coefficient,

$$\nabla H_S \approx nF \nabla \phi_S; \quad \nabla H_e \approx nF \nabla \phi_e \tag{3.39}$$

The current flux density in the electrode and electrolyte phase enters the equation through:

$$i_S = nFN_S; \quad i_e = nFN_e \tag{3.40}$$

Finally we arrive at the energy conservation equation for a lithium ion cell.

$$\rho C_P \frac{\partial T}{\partial t} = \kappa \nabla^2 T - i_S \cdot \nabla \phi_S - i_e \cdot \nabla \phi_e + q_{irr} + q_{rev} + q_{\alpha \to \beta} \tag{3.41}$$

with the heat generation terms, the irreversible, reversible and the contribution due to phase change given respectively by:

$$q_{irr} = a_S F j_i \left(\phi_S - \phi_e - V_0 \right) \tag{3.42}$$

$$q_{rev} = a_S F j_i T \frac{\partial V_{0,p,n}}{\partial T} \tag{3.43}$$

$$q_{\alpha \to \beta} = \Delta h \frac{\partial \left(\frac{r_i}{R} \right)}{\partial t} \tag{3.44}$$

3.5 The Coupled Electrochemical–Thermal Model

Based on the derivations discussed earlier, the conventional expressions of the heat sources are discussed in this section. The equations derived in the last section are repeated and additional design criteria are introduced for better thermal design.

3.5.1 *Thermal Model*

The general energy balance for the lithium ion cell is added to the set of equations, given by Eq. 3.45.

$$\rho C_P \frac{\partial T}{\partial t} = k \frac{\partial^2 T}{\partial x^2} + Q_V. \tag{3.45}$$

In the above equation, the temperature T can vary spatially and temporally, and Q_V is the volumetric source of heat generated during cell operation. The components of the heat source were discussed in the last section. The thermal balance equation is solved with appropriate boundary conditions for the anode, cathode and the electrolyte region. Thus, if δ is the thickness of the cell, the boundary conditions at the anode and cathode current collector ends is given by

$$k \frac{\partial T}{\partial x} = h_0 (T - T_0)|_{x=0,\delta}. \tag{3.46}$$

3.5.2 *The Electrochemical Sources of Heat Generation*

The primary source of heat in the cell is due to the charge transfer reaction, which results in an irreversible and a reversible component for heat generation [4, 19, 29, 33, 35, 39, 49]. The rate of irreversible heat generation due to the charge transfer (Butler-Volmer) reaction, denoted by $Q_{irr,V}$, is obtained by the product of the reaction flux j_k, the surface area per unit volume of electrode a, and the overpotential η. The overpotential is the difference between the solid phase potential ϕ_s, liquid phase potential ϕ_{el} and the equilibrium potential V_0. Rate of irreversible heat per unit volume ($Q_{irr,V}$) is expressed as following and has the units of W m^{-3}.

$$Q_{irr,V} = Faj_n\eta_n; \quad \eta_n = \phi_s - \phi_{el} - V_{0,n}, \tag{3.47}$$

where n corresponds to the positive or negative electrode. This quantity is exothermic both during charge and discharge as the overpotential and the reaction flux change signs between the processes, resulting in irreversible heat generation. The reversible heat generation due to the entropic changes during intercalation, is denoted by $Q_{rev,V}$. This quantity is given by the product of the temperature derivative of the equilibrium potential V_0, reaction flux and the temperature. $Q_{rev,V}$ is expressed in terms of rate of heat release per unit volume and has the units of W m^{-3}.

$$Q_{rev,V} = Faj_kT\frac{\partial V_{0,k}}{\partial T}. \tag{3.48}$$

$\partial V_0/\partial T$ in Eq. 3.48 is related to the entropy [33] of the intercalation reaction

$$\frac{\Delta S_r}{nF} = \frac{\partial V_0}{\partial T}. \tag{3.49}$$

As this quantity is independent of charge or discharge, $Q_{rev,V}$ can be either endothermic or exothermic based on the sign of the rate of reaction, j_k. As the intercalation reaction occurs in the electrode regions, the aforementioned quantities need to be accounted for in positive and negative electrodes and correspondingly $k = p, n$. The secondary contribution comes due to the current conduction in the solid and the electrolyte phases, resulting in the ohmic contribution to heat generation. The ohmic contributions of the individual phases are obtained by the product of current density and the corresponding gradient of the potential [33]. Thus in the electrode region,

$$Q_{ohm,V} = -i_{el}\left(\frac{\partial \phi_{el}}{\partial x}\right) - i_s\left(\frac{\partial \phi_s}{\partial x}\right). \tag{3.50}$$

In the separator region, however, this contribution arises due to the electrolyte only the contribution arising from the first term of Eq. 3.50.

3.5.3 Heat Generation Due to Phase Transition

In the seminal formulation of the general energy balance for lithium ion cells, heat generation due to various sources like reactions, heat capacity effects, phase changes, mixing, electric work and heat transfer to surroundings are considered [49]. In subsequent applications of the model to a general lithium ion cell, many of the contributions are not accounted for [49]. Some positive electrodes are known to undergo phase transition, it is important that the heat generation due to this phenomenon be taken into account. Within the distributed energy source framework, the electrochemical model is extended to incorporate the heat generation due to phase transition and is expressed in terms of the difference in enthalpies of the two phases (α, β) undergoing phase transition, and the rate of phase transition at the $\alpha - \beta$ interface.

$$Q_{pt,V} = \Delta H_{\alpha\beta}\dot{\eta}_{\alpha\beta}. \tag{3.51}$$

Modeling of the heat source rate (Eq. 3.51) is also referred to as equivalent enthalpy approach. Within the framework of the generalized moving boundary model [26], the rate of phase transformation is evaluated as

$$\dot{\eta} = \frac{\dot{r}_i^{\alpha \to \beta}}{r_p} = \frac{1}{r_p} \frac{\mathrm{d}r_i^{\alpha \to \beta}}{\mathrm{d}t}, \tag{3.52}$$

where $r_i^{\alpha \to \beta}$ denotes the interface position. An integral of the rate of heat flow obtained from calorimetric experiments, for the two phase region gives the net enthalpy change for $\alpha - \beta$ transition ($\Delta H_{\alpha\beta}$) and is used in Eq. 3.52. This contribution is accounted only in the positive electrode.

Based on the heat sources discussed above, the total heat generation in the positive electrode is given by the sum of Eqs. 3.47, 3.48, 3.50 and 3.51 and in the negative electrode is the sum of Eqs. 3.47, 3.48 and 3.50. In the separator region, as there is no intercalation reaction, the only contribution is due to $Q_{\mathrm{Ohm},V}$.

3.5.4 Performance Indicators Computed from the Model

While discussing the model equations, the heat sources in a cell during operation are mentioned. In this section the various performance indicators that are computed from the model are introduced. In order to obtain the total rate of heat generation from individual electrodes, each of the volumetric terms mentioned in Eqs. 3.47–3.51 is integrated over the thickness of the corresponding electrode. For example, the rate of irreversible heat generation in the anode is given by

$$Q_{irr,Neg} = \int_0^{\delta_A} Q_{irr,VNeg} dx, \tag{3.53}$$

In a similar manner, the reversible and ohmic contribution from the negative, positive electrodes and separator sections are calculated. The individual total (reversible/irreversible/ohmic) heat generation rate in a cell is obtained by summation of these contributions

$$\begin{aligned} Q_{irr} &= Q_{irr,Neg} + Q_{irr,Pos}, \\ Q_{rev} &= Q_{rev,Neg} + Q_{rev,Pos}, \\ Q_{Ohm} &= Q_{Ohm,Neg} + Q_{Ohm,Pos} + Q_{Ohm,Sep}. \end{aligned} \tag{3.54}$$

The cumulative total of rate of heat generation during discharge is given as:

$$Q_D = Q_{irr} + Q_{rev} + Q_{Ohm} + Q_{pt}. \tag{3.55}$$

It is to be noted that all the heat generation rates mentioned in this section have the units of W m^{-2}. Subsequently, the total amount of heat energy generated, expressed in J m^{-2} can be obtained from the corresponding rates by integration over time. For example, the total cumulative heat generated during discharge is calculated by

$$Q_T = \int_0^t Q_D dt \tag{3.56}$$

Note that Q_T is a function of time and gives the total heat generated at any instant of time during discharge. The value of this quantity at the end of discharge Q_T when $t = t_D$, is the total heat generation for complete discharge. The useful energy output from the cell is the electric energy during discharge. The electric energy is computed by integrating the respective discharge power (VI) over time. For example, the total electric energy output from the cell during discharge is obtained by

$$W_{el,D} = \int_0^{t_D} VI dt. \tag{3.57}$$

It is to be noted that the ratio of $W_{el,D}$ to the thermal energy Q_T is an indicator of the efficiency of the operation.

Chapter 4
Key Applications of Electrochemical Theory

The solution of the electrochemical thermal model developed in the previous chapters is discussed under various charge–discharge scenarios. The total heat generation is resolved to anode, cathode, and separator sections of the Li-ion cell, and the significant sources identified. The contribution due to irreversible, reversible, phase change processes is identified. Dependence of these multiple components on the operating conditions is studied, and optimal operating conditions are suggested.[1]

4.1 Introduction

The characteristic high energy density of lithium-ion battery comes at a cost of self-heating of this power source due to several inherent electrochemical processes. These electrochemical processes can be thermodynamically classified as dissipative and non-dissipative phenomena, which lead to inevitable heating of batteries. Self-heating may lead to severe safety hazards and can lead to phenomenon like thermal run away. It provides the impetus to discern not only the physics of the dissipative thermodynamic processes but also the effect of it on working of these batteries. In order to understand the thermal behavior in detail, it is important to develop the fully coupled electrochemical thermal model. It may be noted that the thermal effects differ across lithium-ion cells of varying chemistries due to the varying electrochemical processes [3]. Thermal stability is the key property that ensures safe operation and efficient thermal management. The electrochemical model is required to obtain information about the individual sources of heat from electrodes/electrolyte and the dependence of the heat generation rates on the operating conditions. Information about the controlling electrode and the optimum operating conditions are important for the design of thermal management systems.

---

[1] Figures and discussions reproduced with permissions from Elsevier.

K. S. Hariharan et al., *Mathematical Modeling of Lithium Batteries*,
Green Energy and Technology, https://doi.org/10.1007/978-3-319-03527-7_4

As discussed in the previous chapters, modeling of the electrochemical processes in a cell has been predominantly based on a macro-homogeneous approach wherein homogeneous equations are used to model the solid matrix and the electrolyte system [10, 11, 14, 34]. Transport in the solid phase is modeled by considering diffusion in the individual particles constituting the electrode resulting in a pseudo-two-dimensional (P2D) model. The generalized structure of the governing partial differential equations (PDE) facilitates incorporation of thermal effects with the electrochemical model. However, complete integration of the thermal model into the P2D has happened in phases. The general energy balance based on thermodynamic principles for an electrochemical system [4] assumes temperature uniformity within the cell. It has been successfully used to incorporate the macro-homogeneous models in a single cell as well as a stack [35]. Later, an alternate approach based on the local heat generation developed subsequently [33, 39] was proven to be equivalent to the thermodynamic approach. A thermal energy equation capable of describing the internal temperature distribution of the cell is developed using the volume-averaging approach [19]. These developments have resulted in various studies where a coupled electrochemical–thermal model is used to analyze the behavior of lithium-ion cells [8, 29, 51]. Thermal coupling enables incorporation of the temperature dependence of thermodynamic [54] and transport properties of the electrochemical model based on experimental observations [55]. These theoretical developments were tracked, with the derivations in the last chapter. In this chapter, we will present some representative results of the electrochemical—thermal model [27]. The reader is suggested to refer many publications in this area for a wider scope of applications. We will focus on the application to Li-ion cells constituted of LFP ($LiFePO_4$) as the positive electrode and MCMB (natural graphite) as the anode. In addition to normal trends in the cell voltage and temperature, these cells exhibit additional features and hence are best suited for demonstration of interplay between the electrochemical and thermal response.

In this work, the intent is to develop the generalized moving boundary model [26] to include the non-isothermal behavior. For modeling electrochemical sources of heat, the contributions from the anode, separator, cathode and phase transformation are incorporated as source terms in the governing energy balance equations. The electrochemical processes are in turn coupled by incorporating the temperature dependences of the mass and charge transport properties [17, 30, 58]. The non-isothermal model thus developed is validated with experimental data. As these cells are used as energy source in various applications, it is important to understand the dependence of heat generation on operating conditions in order to conceive appropriate cooling concepts resulting in efficient battery pack design. In this regard, the dependence of the heat released on the ambient temperature is investigated. In a battery pack, the single cell interacts with the surrounding cooling medium through the heat transfer coefficient. Hence, the dependence of the rate of heat generation on this quantity is studied in detail. The insights obtained from this analysis are used to obtain optimal operating conditions that minimize thermal losses and extend battery life. This study enables to conceive appropriate cooling concepts when cells are used as energy source in automobiles and provide key insights into the design of an efficient battery thermal management system.

In the next section, numerical simulations are performed to understand various aspects of thermal effects in the Li-ion cell considering phase transforming effects in the cathode. The first step toward this involves verification and validation of the model developed. Toward this goal, the model predictions are compared with available experimental results. The validated model is used to numerically analyze the electrochemical sources of heat generation. Further efficacy of this study is brought by drawing useful thermal design conclusions in terms of optimal operating conditions such as ambient temperature, rate of operation, and heat transfer coefficient.

4.2 Validation of Discharge and Charge

In this section, the proposed thermal model is validated by comparing the predicted voltage response with those from experiments during discharge and charge operations. The experimental data considered for validation [13, 17, 30, 58, 59] are from $LiFePO_4$/graphite cells with a capacity of 2.3 Ah. These are cylindrical cells of 26650 type. The model comparison at 273, 298, and 323 K at 1C rate of discharge is shown in Fig. 4.1.

The model compares well with data from experiments, and the error is less than 1% at these temperatures. Typical features of the discharge voltage such as (a) the constant discharge voltage (b) increase in the discharge voltage as well as end-of-discharge capacity with an increase in temperature are predicted accurately by the model. Also, it can be seen that with an increase in the temperature, the end capacity or utilization increases owing to efficient transport at higher temperatures.

Subsequently, the ability of the model to represent high rate operation is tested. Voltage response at 298 K during charging of cylindrical cells at rates ranging from 1C to 10C is reported in the literature [13, 58] and is used for validation of the proposed model. For this validation, the electrolyte diffusivity $\sim 1.3 \times 10^{-}10\ m^2\ S^{-1}$

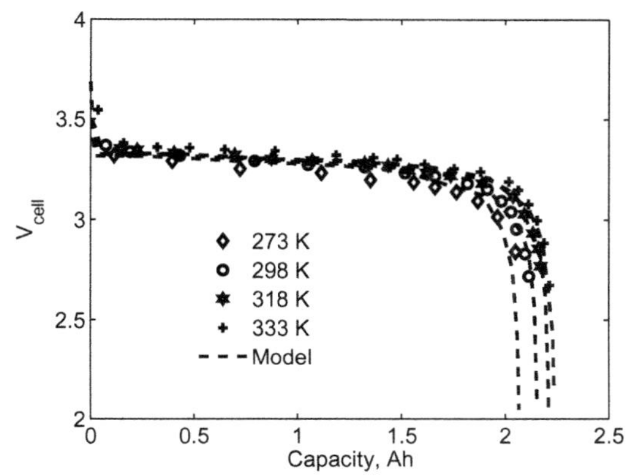

Fig. 4.1 Comparison of voltage results from models and experiments for 1C battery discharge at different ambient temperatures

is used, as suggested [58]. The voltage response of the model and the experiments are compared in Fig. 4.2.

It can be seen that the model prediction of cell voltage matches closely to the results from experiments at all the rates. The charge response of any typical Li-ion

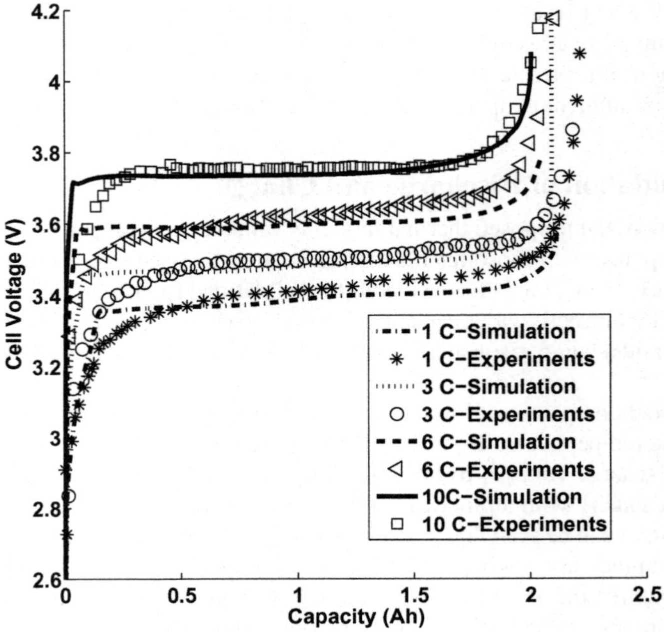

Fig. 4.2 Comparison of voltage results from models and experiments for different charging rates

Fig. 4.3 Comparison of volumetric heat generation from model and experiments for different charging rates

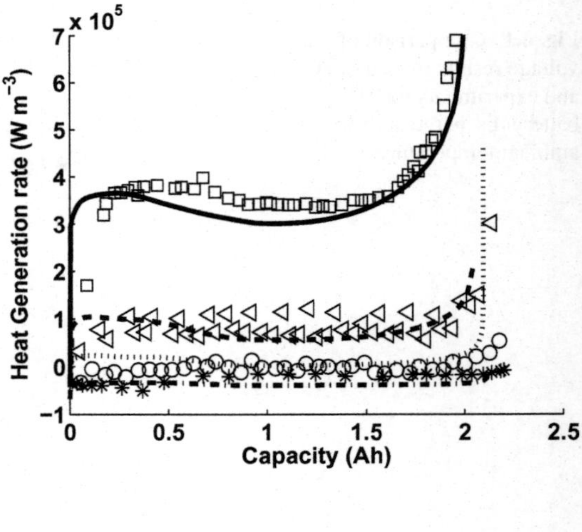

cell is characterized by three regimes: the initial rise, the constant voltage regime, and the steep voltage rise near the end of charge. This nonmonotonic response is attributed to the multiphase behavior of the electrodes [53], and the model captures the significant features at all the rates studied. It may be mentioned here that almost no change in end capacity during charge is seen with the increase in rate. This is due to increased utilization during charging in isothermal condition [50]. Hence, ease in transport due to increase in rate does not play a significant role. The comparison of the model output of heat release with experimental results is very important. The results of this exercise, for charge rates ranging from 1C, 3C, 6C, and 10C, are shown in Fig. 4.3.

For a conventional lithium-ion cell, the heat generation increases with the rate of operation, and sometimes, this quantity increases steeply near the end of charge/ discharge [21]. The model is seen to exhibit both these expected behaviors, and the outputs are close to experiments.

4.3 Analysis of Different Electrochemical Heat Sources

The validated electrochemical thermal model is used to obtain the interplay between different heat sources of Li-ion cell. The contributions from the individual components, at 1C rate and 298 K, are discussed in Fig. 4.4. For the sake of brevity, the sources of heat for the positive (Q_{Pos}) and negative (Q_{Neg}) electrode are shown in the same figure (Fig. 4.4). The individual rates of heat are obtained by integrating the heat generation over the respective electrodes or the separator region, using Eqs. 3.53 and 3.54. The significant contribution to the heat generation in the negative electrode is from the reversible component and is endothermic in nature. The ohmic component and irreversible contributions are comparable for a wide range of discharge capacities. The latter becomes significant near the end of discharge, due to increase in the overpotential at these capacities. In the positive electrode, the reversible component is much larger than the other contributions [22]. The heat release due to phase transition is the next significant contributor. This component, due to the construct described in the earlier sections, contributes only during the two-phase region and is zero during the initial regimes of discharge. The irreversible and the ohmic contributions are much lesser in magnitude in comparison for the positive electrode. Another distinct feature is that the irreversible and reversible components do not significantly change with the discharge capacity. From the analysis of heat sources, it can be concluded that the contribution from the positive electrode (in this case LFP) is lesser in comparison to the negative electrode, and the total heat generation of the cell has signatures of the latter. This observation is in accordance with the reported experimental results [13] of LFP/C cells and is ascribed to the multiphase coexistence. Thus, it can be concluded that by accounting for the phase transition phenomena accurately, the experimental observations can be described better. This analysis also helps to identify the electrode that controls the cell behavior, further aiding in improved cell design. It is to be noted that the significant contribution from either electrode is due to reversible component, which can be recovered during a complete discharge–charge cycle.

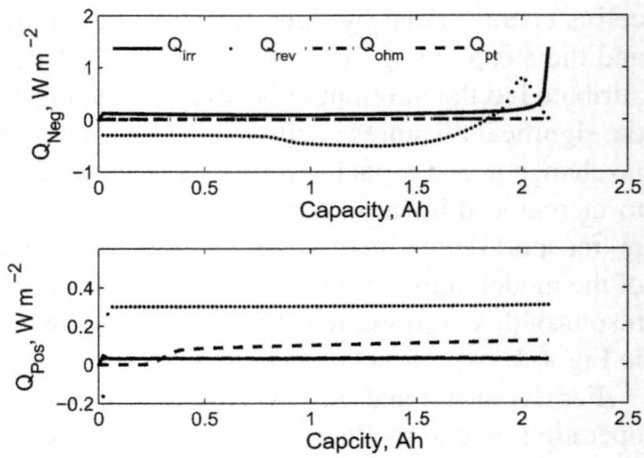

Fig. 4.4 Evolution of spatially averaged different heat source terms with cell capacity for 1C discharge at 298 K

4.4 Effect of Ambient Temperature

The effect of the ambient temperature is discussed subsequently. The degree of self-heating or the rise in temperature when the cell is discharged at a rate of 1C is shown in Fig. 4.5, against the discharge capacity. Here and for all the results reported later, the temperature measured is at the current collector end of the positive electrode. The temperature rise is highest for lower ambient temperatures, consistent with the reported trends [19]. It can also be seen from the figure that the temperature rise is steeper near the end of discharge [21]. This indicates that an optimal choice of the operating range of state of charge (SOC) can be made to reduce the temperature increase substantially. The temperature increase is related to the rate of heat generation from the cell. It is important to identify the most important sources of heat at varying ambient temperatures and is analyzed in the subsequent section. The rates of heat release at various ambient temperatures are shown in Fig. 4.6. In addition to the total heat, contributions from the positive and negative electrodes at various ambient temperatures are shown in Fig. 4.6. In accordance to the results published earlier [19], it can be seen from the figure that the rate of heat generation is higher at lower ambient temperatures. The results indicate that the significant contribution is from the negative electrode, especially nearing the end of discharge. The dominance of negative electrode on heat generation can be owed to the thermal stability of positive electrode. It can be seen from Fig. 4.6 that the heat generated is a constant till capacity of 1 Ah, and it subsequently decreases till around 1.7 Ah, eventually increasing near the end of discharge. The observation that there are regimes where heat generation decreases (between 1 and 1.7 Ah in this case) can be used for efficient thermal design. Thus, it can be seen that by analyzing the heat generation rates, one can choose an optimal temperature and SOC range of operation with minimum thermal losses. On analyzing the results, it can be seen that the decrease in the heat generation is due to the effect of the negative electrode. The heat generated at the negative electrode is exothermic only at 273 K. For the other temperatures, it is endothermic for

Fig. 4.5 Temperature rise or degree of self-heating versus cell capacity for 1C discharge at 298 K

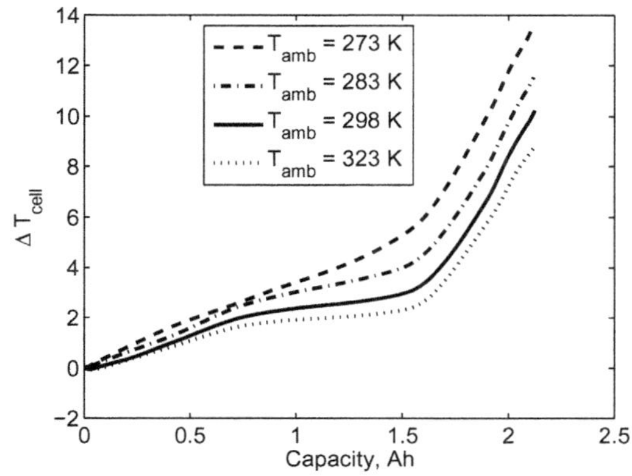

Fig. 4.6 Evolution of spatially averaged total heat generation with cell capacity for 1C discharge at different ambient temperatures (273, 283, 298, and 323 K). Separate plots for cell and two electrodes are shown

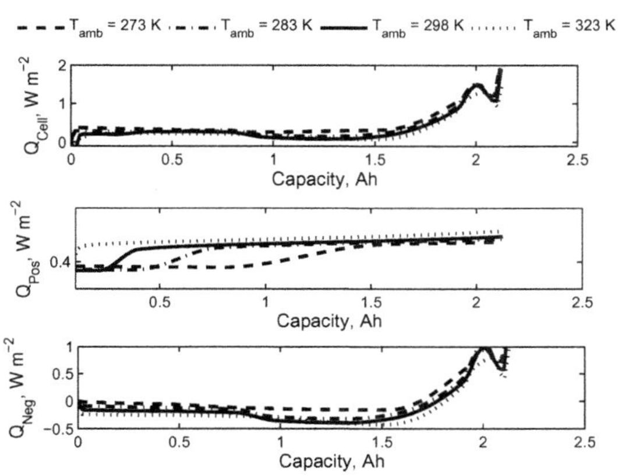

significant regimes of discharge capacities. Comparing the contributions at the corresponding discharge capacities, it can be seen that the regimes at which the negative electrode has endothermic contribution are balanced by an exothermic contribution in the positive electrode. This combination results in the cell that is thermally very stable. Such design tools can be used to screen electrode materials for better thermal stability. Analysis of the positive electrode reveals another interesting scenario. It can be seen from Fig. 4.6 that the rate of heat release is highest at the highest temperature, 323 K. This observation is contrary to the results reported for electrodes with other chemistries. To understand this behavior in detail, and to analyze the individual electrochemical sources of heat, components of the positive electrode are shown in Fig. 4.7 at all the temperatures studied.

It can be seen from Fig. 4.7 that among the components of heat generation, the irreversible component of the charge transfer reaction ($Q_{irr,Pos}$) and the ohmic ($Q_{ohm,Pos}$) components are of comparable magnitude. These two quantities increase with a decrease in the ambient temperature. $Q_{irr,Pos}$ is high at lower temperatures due to higher overpotential at a lower temperature. This feature, in turn, is due to lower val-

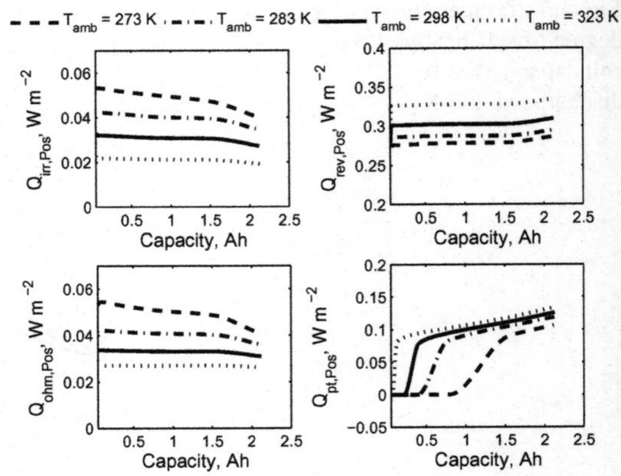

Fig. 4.7 Evolution of spatially averaged different heat source terms of positive electrode with cell capacity for 1C discharge at different ambient temperatures (273, 283, 298, and 323 K)

ues of reaction and transport coefficients. On a subsequent level of analysis, a lower overpotential in LFP arises due to the constant discharge voltage, a characteristic feature in electrodes displaying multiphase coexistence. In comparison, the reversible component is of larger magnitude and is larger at higher ambient temperatures. This is because the reversible component scales linearly with temperature as can be seen in Eq. 3.48. The heat generation due to phase transformation also exhibits a similar trend with operating temperature, because of faster interface movement. Interface movement is coupled to solid phase diffusion transport as given by Eq. 3.52 and is faster at higher temperatures owing to Arrhenius dependence of diffusivity leading to larger heat losses. In scenarios where the significant heat generation is of reversible nature like in the case of an LFP/C cell, the thermal losses are higher at a higher operating temperature. The corresponding contributions at the negative electrode are shown in Fig. 4.8.

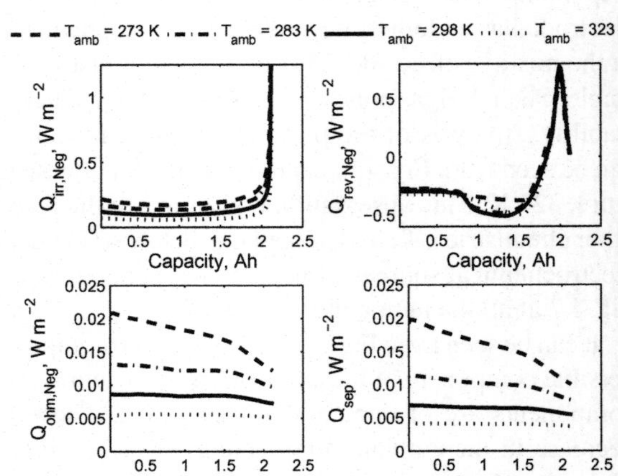

Fig. 4.8 Evolution of spatially averaged different heat source terms of positive electrode with cell capacity for 1C discharge at different ambient temperatures (273, 283, 298, and 323 K)

The significant contributions in the negative electrode, especially near the end of discharge, are due to the irreversible component of the charge transfer reaction and reversible component. The heat generation due to ohmic contributions in the electrode and the separator is lower in comparison. The heat generation increases at lower temperatures, especially for the irreversible and the ohmic components, in accordance with the reported results. A comparison between the ohmic heat contributions from the individual sections of the cell at 1C (Figs. 4.7 and 4.8) reveals that the losses are highest at the positive electrode, followed by the negative electrode and then the separator.

This feature could be due to the lower electronic conductivity of the positive electrode material. Thus, it can be concluded from this section that across ambient temperatures, the overall heat generation is higher at lower temperatures, due to the predominance of the negative electrode contribution. The multiphase coexistence, however, leaves a significant impact on the rates of heat generation.

4.5 Effect of Rate of Discharge

As the lithium-ion battery is the principal energy source in many applications, the cell is subjected to various load conditions. It is important to understand how the heat contributions are affected by discharge rate at various ambient temperatures. In this study, the temperature and heat generation at 1C, 4C, and 8C are analyzed. In order to assess the maximum heat effects, for these studies, the cell is insulated, i.e., the heat transfer coefficient $h = 0\,\mathrm{W\,m^{-2}\,K^{-1}}$. The effect of heat transfer coefficient will be addressed in the next section. The temperature rise of the cell at various ambient temperatures and rates of discharge is shown in Fig. 4.9.

As expected, the temperature rise is largest at the 8C rate for all ambient temperatures. At 8C, the temperature rise is around 40 at 273 K and is close to 30 K

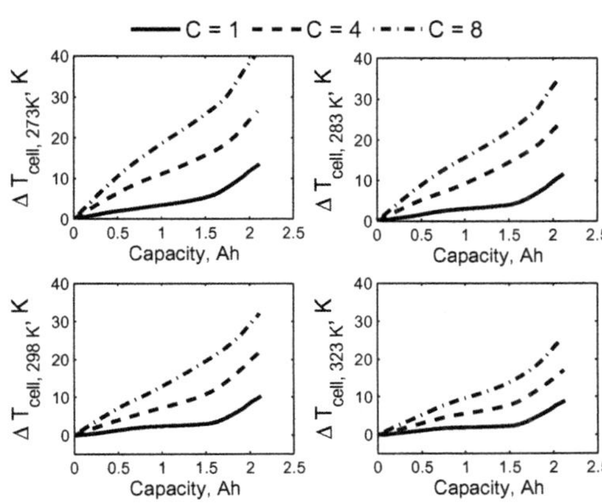

Fig. 4.9 Temperature rise or degree of self-heating versus cell capacity for three different discharge rates 1C, 4C, and 8C for four different ambient temperatures (273, 283, 298, and 323 K)

for 283 and 298 K. Larger temperature rise at 273 K can be ascribed to the higher transport losses at lower temperatures. For 1C rate at all ambient temperatures, the temperature rises gradually till capacity of 2 Ah, which is succeeded by a regime of steep increase. It was mentioned earlier that this is a signature of the multiphase coexistence. At higher rates, this effect is weakened by other factors. To study the effect of discharge rate on the heat generation, results at 8C are analyzed in detail in Fig. 4.10.

At this rate of discharge, the heat generation is exothermic at all temperatures. The heat generation is highest at 273 K and decreases with an increase in the ambient temperature. At higher temperatures (298, 323 K), the cell heat generation is a constant till 1.7 Ah (Depth of Discharge, DOD = 0.74) before a steep rise near the end of discharge. Contrary to the behavior at 1C, the heat generation from the positive electrode is higher at lower temperatures, and both the electrodes and the cell display similar behavior. At the lower temperatures (273 and 283 K), a higher value of Q_{Pos} is seen at initial values of discharge capacity (<0.5 Ah). This feature is reflected in the Q_{Cell} also at the same capacity. This is unlike the 1C case (Fig. 4.6) wherein the positive electrode does not contribute significantly to Q_{Cell}.

To understand the most sensitive component to discharge rate, the individual heat generation from the positive electrode is shown in Fig. 4.11, and the corresponding values of the negative electrode are shown in Fig. 4.12.

In comparison with 1C results (Fig. 4.7), it can be seen that at high rates, the irreversible components ($Q_{irr,Pos}$, $Q_{ohm,Pos}$) of the positive electrode have increased much more than the other components. It is to be noted, however, that even at the high rates, the $Q_{rev,Pos}$ and $Q_{pt,Pos}$ exhibit atypical behavior—it is higher at the highest temperature. It can also be seen from Fig. 4.11 that the phase transition heat is zero for longer ranges of DOD, indicating that the systems exist predominantly in the single phase. The novel features seen for Q_{Pos} in 8C (Fig. 4.10)—higher magnitudes at lower temperatures and lower discharge capacities—can be ascribed to larger irreversible losses due to charge transfer reaction and the ohmic heat (Fig. 4.11). A

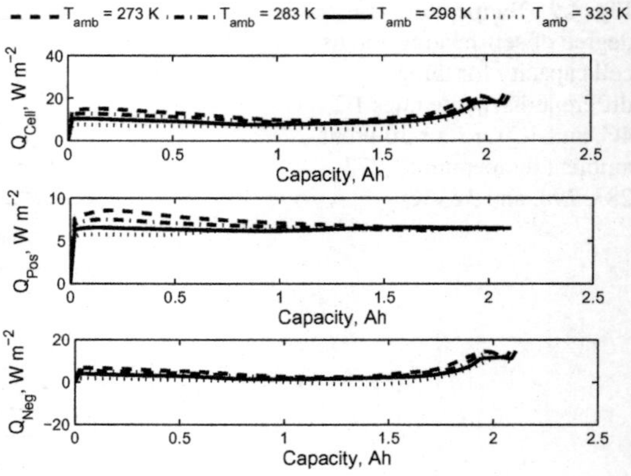

Fig. 4.10 Evolution of spatially averaged different heat source terms with cell capacity for 8C discharge for four different ambient temperatures (273, 283, 298, and 323 K)

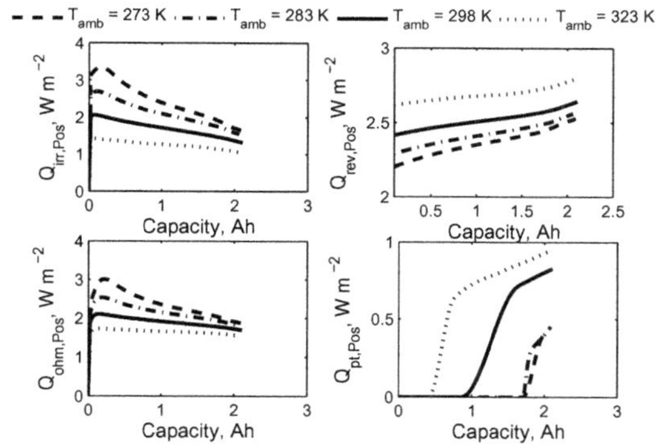

Fig. 4.11 Evolution of spatially averaged different heat source terms of positive electrode with cell capacity for 8C discharge at different ambient temperatures (273, 283, 298, and 323 K)

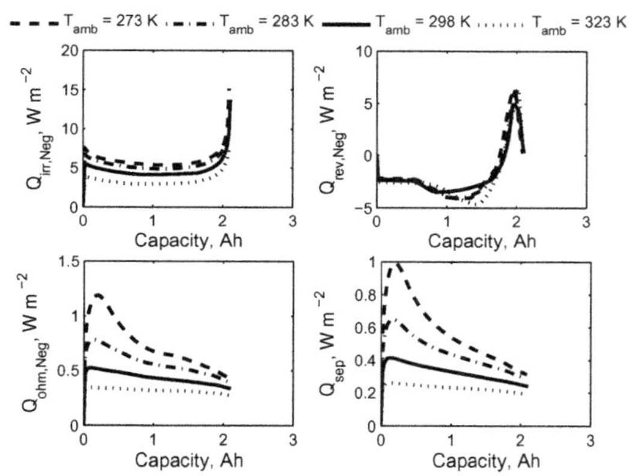

Fig. 4.12 Evolution of spatially averaged different heat source terms of negative electrode with cell capacity for 8C discharge at different ambient temperatures (273, 283, 298, and 323 K)

similar comparison of the heat generation in the negative electrode at 1C (Fig. 4.8) and C (Fig. 4.12) reveals that the relative contribution of the reversible component has reduced at 8C, although a short endothermic region exists at these high rates. Thus, a study of high rate operation reveals that the irreversible components are more significant and that positive electrode has significant heat generation.

4.6 Effect of Heat Transfer Coefficient

The rate of discharge is decided by the driving conditions and cannot be controlled. In order to ensure safe and efficient operation, this heat is removed from the battery pack by the use of a heat transfer medium. The type and mode of operation of the cooling system to ensure minimization of thermal losses and maximization of battery life is the key challenge in battery thermal management. The interaction of the cell with the medium is through the heat transfer coefficient, h, and an efficient thermal design requires choice of an optimal value of this quantity. To enable a thorough understanding from an electrochemical perspective, the effect of h on the heat generation is studied in this section. As can be seen from Fig. 4.9, the temperature rise is most significant at 8C, and the effect on increasing h at this rate is shown in Fig. 4.13.

The values of h used (0, 0.1, 1, and 100 W m^{-2} K^{-1}) roughly correspond to adiabatic, natural convection, forced convection, and liquid cooling scenarios, respectively. It can be seen that at $h = 100$ W m^{-2} K^{-1}, the rise in temperature is negligible while the temperature of the cell saturates with $h = 1$ W m^{-2} K^{-1} indicating safe operation. As the heat generation increases at high rates, the effect of heat transfer coefficient for discharge at 8C is analyzed in Fig. 4.14.

A most significant feature that can be observed from Fig. 4.14 is that the heat generation is higher, at higher values of heat transfer coefficient. Although this effect is significant at 273 and 283 K, this trend is seen for all other ambient temperatures.

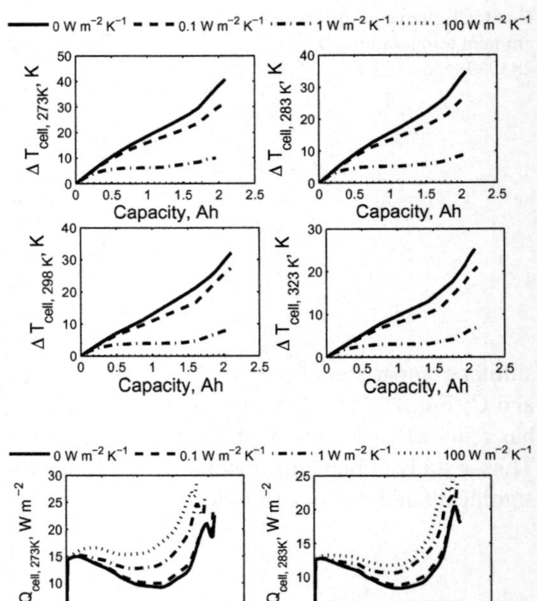

Fig. 4.13 Temperature rise or degree of self-heating versus cell capacity for four different convective film coefficients at the end ($h = 0$, 0.1, 10, 100 W m^{-2} K^{-1}) and for four different ambient temperatures (273, 283, 298, and 323 K)

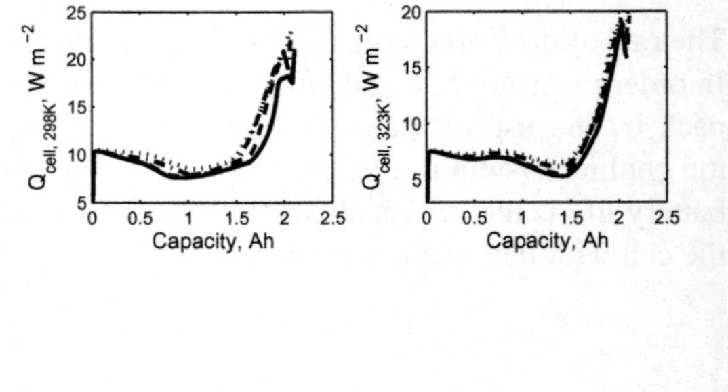

Fig. 4.14 Evolution of spatially averaged different heat source terms versus cell capacity for four different convective film coefficients at the end ($h = 0$, 0.1, 10, 100 W m^{-2} K^{-1} and for four different ambient temperatures (273, 283, 298, and 323 K)

This result implies that the heat generated by the cell at an adiabatic condition is lower than a scenario, where any method of heat removal—natural or forced convention—is used. The electrochemical reason for this nonobvious behavior is explained based on Fig. 4.13.

Higher values of h result in more amount of heat removed from the system, resulting in lower operating temperatures. At lower temperatures, the transport properties and the reaction rates are lower, resulting in higher resistance due to the reaction and diffusion processes. This eventually results in larger overpotentials as can be seen in Fig. 4.15, where the values at the positive and negative electrodes are shown for 273 K. This temperature is chosen as the heat generation is the largest, Fig. 4.10.

The larger overpotentials, in turn, result in larger values of Q_{irr} (Eq. 3.47) in both the electrodes—the significant contributor at high rates—resulting in larger overall heat generation at high rates. A comparison of Figs. 4.13 and 4.14 reveals that although higher values of h lead to lesser temperature rise, it results in higher heat generation. This observation is of significant importance in designing efficient cooling concepts in battery packs. A uniform temperature distribution is preferred for improving the battery life, and for achieving this, a coolant with higher h is preferred, limited typically by cost constraints. Under these conditions, however, higher amounts of the electrical energy taken in by the cell while charging is wasted as heat and is not available as electrical energy during discharge. Thus, a high value of the heat transfer coefficient results in higher thermal losses, and a low value results in higher temperatures affecting battery life. Thus, from this analysis, it can be concluded that optimal design of the cooling system should consider the operating efficiency as well as battery life.

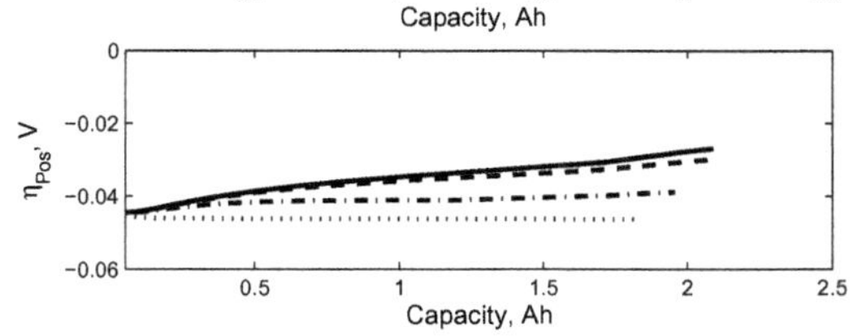

Fig. 4.15 Efficiency versus cell capacity for four different convective film coefficients at the end (h = 0, 0.1, 10, 100 W m^{-2} K^{-1} shown for positive and negative electrode, respectively

4.7 Cumulative Heat Generation and Operating Efficiency Calculation

During battery operation, the two sources of energy are either electrical or thermal. Considering that the thermal energy is a fraction of the total electrical energy the cell has taken up during charging, the total cumulative heat generated from the cell, Eq. 3.56, is a direct indicator of the loss of efficiency during operation. Thus, a detailed electrochemical analysis at various operating conditions helps to identify the optimal conditions, thus minimizing heat losses. The results for the cumulative heat generation at ambient temperatures of 273, 283, 298, and 323 K for adiabatic condition at 1C discharge are shown in Fig. 4.16.

It can be seen that the total heat generation from the positive electrode is exothermic, while it is endothermic from the negative electrode. This results in the total heat generation from the cell being lesser than the positive electrode, indicating a thermally stable scenario. It was mentioned earlier (Fig. 4.6) that the heat generation is highest at 273 K. From Fig. 4.16, it can be seen that the positive electrode shows the reverse trend. Thus, it can be concluded that the endothermic component, due to the reversible part, stabilizes the cell.

Based on the concept of cumulative heat generation, an operating efficiency can be defined. The ratio of the cumulative heat generated (Eq. 3.56) to the electrical work generated from the cell (Eq. 3.57) gives an estimate of the energy that is wasted from the cell. This ratio estimates the amount of heat generated in Joules, for every Joule of electrical work done by the cell. Based on this measure, a better design of lithium-ion battery can be obtained for efficient thermal management by minimizing this factor. This ratio is computed for all the state points that include ambient temperatures, heat transfer coefficients, and rates of operation, and is reported in percentage in Fig. 4.17.

From the analysis of Fig. 4.17, it can be seen that for 273 and 283 K, using a high h at 8C, results in inefficient operation, as around 6–8 % of the useful work is lost as heat. There is no significant improvement under adiabatic condition compared with

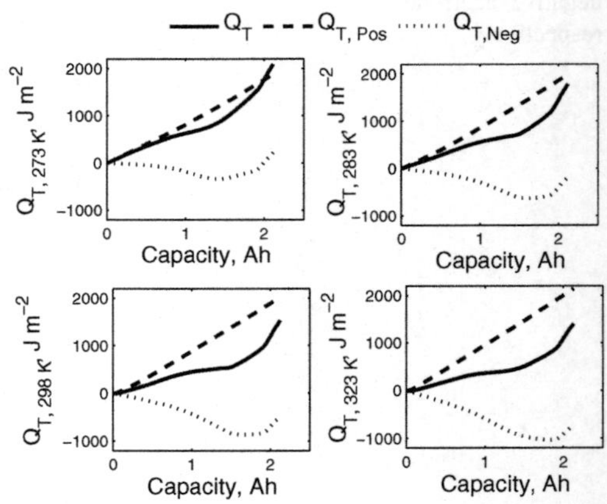

Fig. 4.16 Total cell heat generation for different ambient temperatures (273, 283, 298, and 323 K) under 1C discharge condition for $h = 0\ \mathrm{W\ m^{-2}\ K^{-1}}$

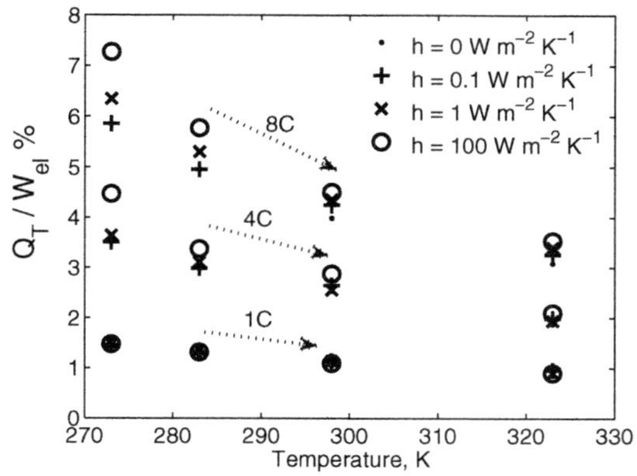

Fig. 4.17 Cumulative heat generation for four different film coefficients (h = 0, 0.1, 10, 100 W m^{-2} K^{-1}) and three different discharge rates 1C, 4C, and 8C, respectively

the natural convection (h = 0.1 W m^{-2} K^{-1}) either. At 4C, the losses are about 4, and 2% at 1C. A judicious choice of the cooling methodology can be made by considering whether the rise in temperature (Fig. 4.13) as well as the thermal losses (Fig. 4.17) is within agreeable limits. Given a 5 K rise in temperature is within safe operating limits, choice between a forced air convection (h = 1 W m^{-2} K^{-1}) or liquid cooling (h = 100 W m^{-2} K^{-1}) can be made at the cost of a heat loss. It can also be inferred from this section that a simple feedback control, which increases the coolant flow rate to compensate for temperature increase, may not be a most efficient method, as this would result in larger heat generation. An efficient control system should also consider the heat losses in addition to the temperature, to obtain an optimal cooling scenario. From the results for the single cell, the thermal losses seem to account for $<8\%$ of the electrical work. In a realistic scenario of a battery pack, however, this ratio could be much higher due to all the other components of the battery pack.

4.8 Summary

In this part, we derive the electrochemical modeling framework for lithium-ion cells. We start from simple equations that relate fluxes and forces, and develop it stepwise to obtain the conservation equations for mass and charge in the electrolyte phase. Corresponding equations for the electrode phase are obtained by direct application of the basic rules of transport phenomena, Fick's and Ohm's laws, respectively. The model for charge transfer reaction is derived from thermodynamics. In the second chapter, the energy balance for lithium-ion cells is obtained from essential thermodynamics. The evolution from a lumped to a distributed heat source is covered after a brief foray into nonequilibrium thermodynamics. The outcome is the celebrated and much successful pseudo-two-dimensional (P2D) model for lithium-ion cells.

In the last chapter, typical applications of the electrochemical thermal model are illustrated. The value add is the design guidelines that come out of the analysis of the heat sources. The significance of the thermal electrochemical coupling is demonstrated by a systematic study of the thermal response of individual components of a Li-ion cell. It is observed that the biphasic nature of electrode material imparts an atypical thermal response of higher heat generation with increasing ambient temperatures. At high rates of operation, the contribution to the heat generation is dominated by the irreversible and ohmic sources of heat. It is observed that high values of heat transfer coefficient lead to lower values of cell temperature but result in higher heat generation. This insight is of importance in designing an efficient thermal management system. Using the cumulative heat generation as a measure of the operating efficiency, it is demonstrated that using a cooling medium with a high heat transfer coefficient can lead to inefficiency in the cell operation at high rates.

Part II
Reduced Order Models

Chapter 5
Theoretical Framework of the Reduced Order Models (ROM)

Physically motivated model order reduction of the electrochemical–thermal model is developed. The motivation is to enable the use of the ROM for on-board scenarios. The equations are lighter to solve and are intuitive, with parameters retaining the physical relevance.[1]

5.1 Introduction

In the earlier chapters, we discussed the electrochemical processes in a Li-ion cell, and also how the underlying physics can be expressed as mathematical equations. The ensuing construct is a set of ten coupled partial differential equations (PDEs). As we know, electrodes contain a solid phase comprising of active materials, binders, and the electrolyte phase. Separator typically has some solid polymeric membrane and a liquid electrolyte phase. Thus, there are totally five phases: active materials in the two electrodes and electrolyte phase in the three regions. Li ion gets shuttled between the electrodes, carrying its own mass, and a positive charge with it. Both these quantities are conserved—mass and charge balances for these five phases resulting in 10 equations. A PDE is solved by a finite difference or a more sophisticated finite element method. The latter requires specialist solvers to do the task, and there has been no report of an onboard application. The simpler finite difference method can be solved in an implicit or an explicit method. Latter though simple, can be unstable, and the former requires multiple iterations for each time step. The suit of both these approaches for conditions of normal use of a battery—either with multiple apps in a mobile or aggressive driving conditions in an EV, is not clear. In addition, the battery pack of an electric vehicle has hundreds of cells, each in various states. It is impossible to run hundreds of PDE solvers, and a single model could be too simplistic. From a cell design perspective, PDE solution also hampers cycle life predictions, as the simulation timescales up to months. A PDE-based solution framework makes it difficult—if not impossible—to have a state estimation algorithm

[1]This approach was first developed in [47]

K. S. Hariharan et al., *Mathematical Modeling of Lithium Batteries*,
Green Energy and Technology, https://doi.org/10.1007/978-3-319-03527-7_5

to run in tandem with the base equation set. The first step of simplification is to obtain the corresponding manageable set of ordinary differential equations (ODEs). In order to retain the physical basis of the equations the first challenge would be to identify the accurate corresponding variable, the time dependence of which encompasses the time and space dependences of all the variables discussed in previous chapters. A list of the governing equations and the corresponding boundary conditions involved in EM are provided in Table 5.1. Volume averaging is a neat trick that can be used in this scenario [47]. As the quantity is already averaged over a given volume, the spatial dependence naturally drops off, and we are bothered about the time evolution only. A formal mathematical definition is for the average of any system variable $f(x, t)$ across a region of thickness L with a surface area S, in the Cartesian coordinates is defined as follows:

$$\langle f(t) \rangle = \frac{1}{S \times L} \int_0^L f(x, t) S dx, \tag{5.1}$$

where $\langle\rangle$ denotes the volume averaged quantity. The average of a variable in the radial coordinates (denoted by an overbar) for a particle of radius R is defined as follows:

$$\bar{f}(t) = \frac{1}{\frac{4}{3}\pi R^3} \int_0^R 4\pi r^2 f(r) dr, \tag{5.2}$$

For example, the volume average reaction rate in the negative electrode is as follows:

$$\langle j_n(t) \rangle = \frac{1}{S \times L} \int_0^{L_n} j_n(x, t) dx, \tag{5.3}$$

A field equation is a volume averaged by integrating it over its domain, and normalizing with the domain volume. We can use this volume averaging principle to reduce the original PDEs to ODEs, and even reduce the original ECT model all the way up to a set of algebraic equations [48].

Battery is a funny system, as all the relevant action happens at the interfaces. The primary output from the model is the cell voltage. The cell voltage is defined as the potential drop between the cathode current collector and the anode current collector. Formally,

$$V_{\text{cell}} = \phi_{sp}|_{x=L} - \phi_{sn}|_{x=0}. \tag{5.4}$$

Similarly, the charge transfer that releases the Li ions and electrons, happens at the interface of electrolyte and the surface of the active material. Hence, a description with only volume-averaged quantities would not work and would need to be complimented by some description of the spatial profile. The latter often starts off as an assumption and verified against the solution of the detailed PDE equation. The profiles are so chosen that they respect basic rules like symmetry of the problem. In the next sections, we will look at various approaches in the model order reduction of

Table 5.1 A compilation of governing equations and the boundary conditions used in the complete electrochemical model

Solid phase mass balance		
$\frac{\partial c_{1k}}{\partial t} = \frac{1}{r^2}\frac{\partial}{\partial r}\left(r^2 D_{1k}\frac{\partial c_{1k}}{\partial r}\right)$ where $k = n, p$		
Boundary Conditions		
$-D_{1k}\frac{\partial c_{1k}}{\partial r}\vert_{r=R_k} = j_k, -D_{1k}\frac{\partial c_{1k}}{\partial r}\vert_{r=0} = 0$		
$c_{1k}(r,0) = c_{1k0}$		
Solid phase charge balance		
$\frac{\partial}{\partial x}\left(-\sigma_{1k}\frac{\partial \phi_{1k}}{\partial x}\right) = -a_k F j_k$		
Negative electrode	Positive electrode	
$-\sigma_{1n}\frac{\partial \phi_{1n}}{\partial x}\vert_{x=0} = I(t)$	$-\sigma_{1p}\frac{\partial \phi_{1p}}{\partial x}\vert_{x=L_n+L_s} = 0$	
$-\sigma_{1n}\frac{\partial \phi_{1n}}{\partial x}\vert_{x=L_n} = 0$	$-\sigma_{1p}\frac{\partial \phi_{1p}}{\partial x}\vert_{x=L} = I(t)$	
Electrolyte phase mass balance		
$\varepsilon_{2k}\frac{\partial c_{2k}}{\partial t} = \frac{\partial}{\partial x}\left(D_{2n}\frac{\partial c_{2k}}{\partial x}\right) + a_n(1-t_+)j_k$		
Negative electrode	Positive electrode	Separator
$-D_{2n}\frac{\partial c_{2n}}{\partial x}\vert_{x=0} = 0$	$-D_{2p}\frac{\partial c_{2p}}{\partial x}\vert_{x=L_n+L_s} = q_{2ip}$	$-D_{2s}\frac{\partial c_{2s}}{\partial x}\vert_{x=L_n} = q_{2in}$
$-D_{2n}\frac{\partial c_{2n}}{\partial x}\vert_{x=L_n} = q_{2in}$	$-D_{2p}\frac{\partial c_{2p}}{\partial x}\vert_{x=L} = 0$	$-D_{2s}\frac{\partial c_{2s}}{\partial x}\vert_{x=L_n+Ls} = q_{2ip}$
$c_{2n}\vert_{x=L_n} = c_{2in}$	$c_{2p}\vert_{x=L_n+L_s} = c_{2ip}$	$c_{2s}\vert_{x=L_n} = c_{2in}$
		$c_{2s}\vert_{x=L_n+L_s} = c_{2ip}$
Total charge balance (solid + electrolyte)		
$-\sigma_{1k}\frac{\partial \phi_{1k}}{\partial x} - \kappa_{1k}\frac{\partial \phi_{2k}}{\partial x} + 2\kappa_{2n}\frac{R_g T}{F}(1-t_+)\frac{\partial \ln c_2}{\partial x} = I$		
Negative electrode	Positive electrode	Separator
$-\kappa_{2n}\frac{\partial \phi_{2n}}{\partial x}\vert_{x=0} = 0$	$-\kappa_{2p}\frac{\partial \phi_{2p}}{\partial x}\vert_{x=L_n+L_s} = I(t)$	$-\kappa_{2s}\frac{\partial \phi_{2s}}{\partial x}\vert_{x=L_n} = I(t)$
$-\kappa_{2n}\frac{\partial \phi_{2n}}{\partial x}\vert_{x=L_n} = I(t)$	$-\kappa_{2p}\frac{\partial \phi_{2p}}{\partial x}\vert_{x=L} = 0$	$-\kappa_{2s}\frac{\partial \phi_{2s}}{\partial x}\vert_{x=L_n+Ls} = I(t)$
$\phi_{2n}\vert_{x=L_n} = \phi_{2in}(t)$	$\phi_{2p}\vert_{x=L_n+L_s} = \phi_{2ip}(t)$	$\phi_{2s}\vert_{x=L_n} = \phi_{2in}(t)$
		$\phi_{2s}\vert_{x=L_n+L_s} = \phi_{2ip}(t)$

the ECT that enables on board/device usage. As we saw in earlier chapters, Li-ion cell has two length scales: the smallest is the active materials (assumed as spherical particles) in the electrodes, and the other is the electrode dimension. We will start the discussion with the smallest first and graduate to the cell level. Subsequently, we will present further developments to this basic reduced order model (ROM). In particular, we will first demonstrate coupling of the thermal model with the ROM [16]. We will end this chapter with a discussion on modeling the phase change exhibited by the positive electrodes [15].

5.2 Solid Phase Mass Balance Equations: A Single Sphere

Conventionally the mass transfer equation in the solid phase and the diffusion equation in the radial coordinates are solved using this approach. The simplest of the solution is called as a single particle model. Let us work through the steps in the next subsection. As the diffusion equation is a second order equation, the general approach would be as follows:

1. Start with the mass conservation equation. This equation is first order in time corresponding to accumulation. This quantity is balanced by the gradient of flux—second order in space.
2. Volume average of the conservation equation results in an ODE for time evolution equation for the averaged concentration. This quantity is balanced by the flux at the surface.
3. The gradient of flux is assumed to have a particular spatial profile, with unknown time-dependent constants.
4. The concentration profile expressed in terms of the constants.
5. The unknown constants are determined by successive integration and volume averaging.

5.2.1 The Diffusion Equation in Radial Coordinates and the Volume Averaged Version

Volume averaging of a function f(r) within a sphere is defined as follows:

$$\bar{f} = \frac{1}{\frac{4}{3}\pi r_k^3} \int_{r=0}^{r_k} 4\pi r^2 f(r)dr = \frac{3}{r_k^3} \int_{r=0}^{r_k} r^2 f(r)dr \tag{5.5}$$

The solid phase mass balance is just the diffusion equation in the radial coordinates. The particles of the active materials are assumed as spheres, and the Li^+ that has received (or about to give up) an electron at the surface, diffuses in and out of these spheres. Thus the flux at the surface is the rate of the Butler–Volmer reaction. Due to radial symmetry the sphere diffusive flux is zero at the center. The system of equations given below is identical for the two electrodes.

$$\frac{\partial c_{sk}}{\partial t} = \frac{1}{r^2}\frac{\partial}{\partial r}\left(r^2 D_{sk}\frac{\partial c_{sk}}{\partial r}\right); k = n, p \tag{5.6}$$

with the boundary and initial conditions as follows:

$$-D_{sk}\frac{\partial c_{sk}}{\partial r}|_{r=r_k} = j_k; -D_{sk}\frac{\partial c_{sk}}{\partial r}|_{r=0} = 0; c_{sk}(r, 0) = c_{sk0}. \tag{5.7}$$

Sphere volume averaging the left hand side of Eq. 5.6 shows (as the volume averaging over the radius and time derivative are independent operations):

$$LHS = \frac{3}{r_k^3}\int_{r=0}^{r_k} r^2 \frac{\partial c_{sk}}{\partial t} dr = \frac{\partial}{\partial t}\frac{3}{r_k^3}\int_{r=0}^{r_k} r^2 c_{sk} dr = \frac{d\bar{c}_{sk}}{dt} \tag{5.8}$$

The right hand side of Eq. 5.6 is sphere volume averaged, using the boundary conditions Eq. 5.7:

$$RHS = \overline{\frac{1}{r^2}\frac{\partial}{\partial r}\left(r^2 D_{sk}\frac{\partial c_{sk}}{\partial r}\right)} = \frac{3}{r_k^3}\int_{r=0}^{r_k} \frac{1}{r^2}\frac{\partial}{\partial r}\left(r^2 D_{sk}\frac{\partial c_{sk}}{\partial r}\right) r^2 dr = \left(r^2 D_{sk}\frac{\partial c_{sk}}{\partial r}\right)_{r=0}^{r_k} = -3\frac{j_k}{r_k} \tag{5.9}$$

Thus the sphere volume averaged solid state diffusion equation is as follows:

$$\frac{d\bar{c}_{sk}}{dt} = -3\frac{j_k}{r_k} \tag{5.10}$$

Equation 5.10 demonstrates that on volume averaging, the surface flux appears in the evolution equation.

From the original initial condition (5.7), the initial condition for the above ODE is obtained as follows:

$$\bar{c}_{sk0} = c_{sk0} \tag{5.11}$$

5.2.2 *Profile Approximations*

The following simple results of volume averaging are needed in this section. These can be obtained from the definition of volume averaging.

$$\bar{r} = \frac{3}{4}r_k;\ \bar{r^2} = \frac{3}{5}r_k^2;\ \bar{r^3} = \frac{1}{2}r_k^3;\ \bar{r^4} = \frac{3}{7}r_k^4 \tag{5.12}$$

As discussed before, the gradient of diffusive flux is assumed to be parabolic in the radial coordinate.

$$\frac{1}{r^2}\frac{\partial}{\partial r}\left(r^2 D_{sk}\frac{\partial c_{sk}}{\partial r}\right) = G(t) + H(t)\frac{r^2}{r_k^2} \tag{5.13}$$

Sphere volume averaging Eq. 5.13 using the above result gives Eq. 5.12

$$\overline{\frac{1}{r^2}\frac{\partial}{\partial r}\left(r^2 D_{sk}\frac{\partial c_{sk}}{\partial r}\right)} = G + \frac{3}{5}H = -3\frac{j_k}{r_k} \tag{5.14}$$

This gives one equation for the two unknowns G and H. Rearranging Eq. 5.14 and integrating once with respect to r and applying symmetry boundary condition and rearrangement gives

$$\frac{\partial c_{sk}}{\partial r} = \frac{G}{3D_{sk}} r + \frac{H}{5r_k^2 D_{sk}} r^3 \tag{5.15}$$

Define:

$$\bar{c}_{skr} = \overline{\frac{\partial c_{sk}}{\partial r}} \tag{5.16}$$

Volume averaging Eq. 5.15 and using 5.12 results in:

$$\frac{4D_{sk}}{r_k}\bar{c}_{skr} = G + \frac{2}{5}H \tag{5.17}$$

This is a second equation for the two unknowns G and H. Equations 5.14 and 5.17 are solved simultaneously to get explicit expressions for G and H as follows:

$$G = \frac{6}{r_k} j_k + \frac{12D_{sk}}{r_k}\bar{c}_{skr};\ H = -\frac{15}{r_k} j_k - \frac{20D_{sk}}{r_k}\bar{c}_{skr} \tag{5.18}$$

Next, an evolution equation for the spherical average radial gradient $\bar{c}_{skr}$ is derived. Differentiating the solid state diffusion equation with respect to r and using the equality of mixed derivatives gives

$$\frac{\partial}{\partial r}\left(\frac{\partial c_{sk}}{\partial t}\right) = \frac{\partial}{\partial t}\left(\frac{\partial c_{sk}}{\partial r}\right) = \frac{\partial}{\partial r}\left[\frac{1}{r^2}\frac{\partial}{\partial r}\left(r^2 D_{sk}\frac{\partial c_{sk}}{\partial r}\right)\right] \tag{5.19}$$

Using the profile approximation Eq. 5.13 for the radial diffusive flux gives

$$\frac{\partial}{\partial t}\left(\frac{\partial c_{sk}}{\partial r}\right) = H\frac{2r}{r_k^2} \tag{5.20}$$

Sphere volume averaging the above equation, using Eq. 5.12 for the averaging of the linear term on the right hand side, and simplifying gives the following equation:

$$\frac{d\bar{c}_{skr}}{dt} = \frac{3}{2}\frac{H}{r_k} \tag{5.21}$$

Substituting for H from Eq. 5.18 and simplifying gives the evolution equations for the average radial gradients as follows:

$$\frac{d\bar{c}_{skr}}{dt} + \frac{30D_{sk}}{r_k^2}\bar{c}_{skr} = -\frac{45}{2r_k^2} j_k \tag{5.22}$$

These ODEs for the average radial gradient are solved with the initial conditions, when the initial states are in equilibrium, with no concentration gradients.

$$\bar{c}_{skr}(r, 0) = 0. \tag{5.23}$$

Finally, the concentration profile can be obtained by integrating Eq. 5.15 once more with respect to r gives solution for the concentration profile, and applying the surface concentration condition $c_{sur,k}$ to eliminate the integration constant as follows:

$$c_{sk} = c_{sur,k} - \frac{G}{6D_{sk}}\left(r_k^2 - r^2\right) - \frac{H}{20D_{sk}r_k^2}\left(r_k^4 - r^4\right) \tag{5.24}$$

Sphere volume averaging Eq. 5.24 and rearranging gives

$$c_{sur,k} = \bar{c}_{sk} + \frac{r_k^2}{15D_{sk}}G + \frac{r_k^2}{35D_{sk}}H \tag{5.25}$$

Substituting the expression for G and H, Eq. 5.18 in the above equation and simplifying gives

$$c_{sur,k} = \bar{c}_{sk} - \frac{r_k}{35D_{sk}}j_k + \frac{8r_k}{35}\bar{c}_{skr} \tag{5.26}$$

5.2.3 Building the Battery Model

The direct reason we need the $c_{sur,k}$ is that the electrode potentials of the electrodes depend on this quantity (Please recollect 2.85, and the section on charge transfer reaction of Chap. 2 in general). Now, we are at a level where we are able to make the simplest model for the cell voltage. At equilibrium, when the current is zero and no transients, the cell voltage is the OCV, which in turn is the difference between the open circuit potentials of the positive and negative electrodes.

$$V_{\text{cell}} = V_{0,cell} = V_{0,p}\left(c_{sur,p}\right) - V_{0,n}\left(c_{sur,n}\right). \tag{5.27}$$

However, as we know from earlier chapters this is only one component that accounts for the cell voltage. During normal operation, voltage drops due to Li-ion transport—electrolyte/electrode conduction, and reaction—diffusion also result in voltage drop.

5.2.4 Butler–Volmer Kinetics

Butler–Volmer kinetics for the electrochemical reactions occurring on the surface of the active material spheres in the electrodes are the following:

$$j_k = 2j_{k0}\sinh\left[\frac{F}{2R_gT}\left(\phi_s - \phi_l - V_{0k}\right)\right]; k = n, p. \tag{5.28}$$

where

The reaction rate prefactor (j_{k_0}) given by

$$j_{k0} = K_k \left(c_{s_{maxk}} - c_{sur,k}\right)^{0.5} c_{sur,k}^{0.5} c_{lk}^{0.5} \tag{5.29}$$

where K denotes the reaction rate, $c_{s_{maxk}}$ is the maximum concentration within the solid phase, c_{sk} is the surface concentration and c_{lk} is the electrolyte phase concentrations and V_{0k} are the electrode open circuit potentials. Inverting Eq. 5.28 and rearranging gives

$$\phi_{sk} = V_{0,k} + \phi_l + \frac{2R_gT}{F}\sinh^{-1}\left(\frac{j_k}{2j_{k0}}\right) \tag{5.30}$$

Using Eq. 5.30 at both the ends of the cell in Eq. 5.4, the cell voltage equation becomes

$$\begin{aligned} V_{\text{cell}} &= V_{0,p}\left(c_{sur,p}\right)|_{x=L} - V_{0,n}\left(c_{sur,n}\right)|_{x=0} + \phi_{l,x=L} - \phi_{l,x=0} \\ &+ \frac{2R_gT}{F}\sinh^{-1}\left(\frac{j_p}{2j_{p0}}\right)|_{x=L} - \frac{2R_gT}{F}\sinh^{-1}\left(\frac{j_n}{2j_{n0}}\right)|_{x=0} \end{aligned} \tag{5.31}$$

There are limitations in this model, namely,

- To use this equation, the reaction rate needs to be known in closed form
- The reaction rate is a local value, a way to tie this up with a measured variable is required
- Voltage due to electrolyte potential is ignored completely $\phi_{l,x=L} - \phi_{l,x=0}$, as no information is available.
- Voltage due to electronic conduction is also ignored.

We need to note that the transport processes occur throughout the length of the cell, and are not local. We just need to apply our knowledge of volume averaging to other equations across the cell domain and keep improving the model. The reaction rate and the sphere average concentration in the above equation are functions of x, in general. Volume averaging across the electrode volume, using definitions of volume averaging, solid state diffusion equation at the electrode level, Eq. 5.10 gives

$$\frac{d\langle\bar{c}_{sk}\rangle}{dt} = -3\frac{\langle j_k\rangle}{r_k} \tag{5.32}$$

We still do not know anything about the volume-averaged reaction rate across the electrode. The next section gives a key input in this direction.

5.3 Solid Phase Current Balance Equations

5.3.1 Field Equations

The generalized Ohms law as discussed in Chap. 2, describes the relationship between current density and potential field in a conductor. The gradient of current density within the electrodes is proportional to electrochemical reaction rate. These equations are the solid phase current balance equations in the two electrodes, given below along with boundary conditions. At the two collector ends, solid current density equals the total current density, computed as external current per unit electrode cross-sectional area. At the separator interfaces, solid current is zero, so that within the separator the current is fully ionic.

5.3.2 Negative and Positive Electrode Regions

As a recap, the solid phase charge balance equation with the boundary conditions is given by

$$\frac{\partial}{\partial x}\left(-\sigma_{1k}\frac{\partial \phi_{1k}}{\partial x}\right) = -a_k F j_k \tag{5.33}$$

$$\sigma_{1n}\frac{\partial \phi_{1n}}{\partial x}|_{x=0} = I(t);\ -\sigma_{1p}\frac{\partial \phi_{1p}}{\partial x}|_{x=L_n+L_s} = 0, \tag{5.34}$$

$$-\sigma_{1n}\frac{\partial \phi_{1n}}{\partial x}|_{x=L_n} = 0;\ -\sigma_{1p}\frac{\partial \phi_{1p}}{\partial x}|_{x=L} = I(t) \tag{5.35}$$

Volume averaging the solid phase charge balance equation in the negative electrode results in the following:

$$\frac{1}{L_n}\int_0^{L_n} d\left(-\sigma_{sn}\frac{\partial \phi_s}{\partial x}\right) = -a_n F\frac{1}{L_n}\int_0^{L_n} j_n dx = \frac{1}{L_n}\left(-\sigma_{sn}\frac{\partial \phi_s}{\partial x}\right)_{x=0}^{L_n} = -a_n F\langle j_n\rangle \tag{5.36}$$

Applying the boundary conditions Eq. 5.34 as the lower and upper limits of integration in Eq. 5.36 gives

$$I = a_n L_n F\langle j_n\rangle \tag{5.37}$$

Similarly volume averaging for the positive electrode gives

$$I = -a_p L_p F\langle j_p\rangle \tag{5.38}$$

Equations 5.37 and 5.38 can be rearranged to get an expression for the average reactions rate in the negative electrode is as follows:

$$\langle j_n \rangle = \frac{I}{a_n L_n F} \tag{5.39}$$

and positive electrode is as follows:

$$\langle j_p \rangle = -\frac{I}{a_p L_p F} \tag{5.40}$$

Equations 5.39 and 5.40 relate the volume average reaction rates in the electrodes to the external current. Please note that this relation holds at any time, for any time varying current profile. These are exact results without any approximations.

Equating 5.39 to 5.40 and rearranging gives

$$a_n L_n F \langle j_n \rangle + a_p L_p F \langle j_p \rangle = 0 \tag{5.41}$$

that retrieves electro-neutrality. In a closed circuit, the electrons generated in an electrode are consumed in the other electrode, with no loss in the external circuit.

5.3.3 Revisiting the Battery Model

Armed with the closure for the reaction rate in terms of the measurable external current (Eqs. 5.39 and 5.40), we can go back to our volume-averaged diffusion equation that was developed earlier, and solve for the concentration profile and hence the cell voltage as follows:

1. For the average concentration in the negative electrode:

$$\frac{d\langle \bar{c}_{sn} \rangle}{dt} = -3\frac{I}{a_n L_n F r_n} \tag{5.42}$$

2. and for positive electrode:

$$\frac{d\langle \bar{c}_{sp} \rangle}{dt} = 3\frac{I}{a_p L_p F r_p} \tag{5.43}$$

3. The new result can be applied to the gradient of average concentration resulting in the following equation for negative:

$$\frac{d\langle \bar{c}_{snr} \rangle}{dt} + \frac{30 D_{sn}}{r_k^2}\bar{c}_{snr} = -\frac{45}{2r_n^2}\frac{I}{a_n L_n F} \tag{5.44}$$

4. for the positive electrode:

$$\frac{d\langle \bar{c}_{spr}\rangle}{dt} + \frac{30D_{sn}}{r_k^2}\bar{c}_{spr} = \frac{45}{2r_p^2}\frac{I}{a_pL_pF} \tag{5.45}$$

5. Finally, the surface concentration of negative and positive electrodes:

$$c_{sur,n} = \bar{c}_{sn} - \frac{r_n}{35D_{sn}}\frac{I}{a_nL_nF} + \frac{8r_n}{35}\bar{c}_{snr} \tag{5.46}$$

6. positive electrode:

$$c_{sur,p} = \bar{c}_{sp} + \frac{r_p}{35D_{sp}}\frac{I}{a_pL_pF} + \frac{8r_p}{35}\bar{c}_{spr} \tag{5.47}$$

7. By replacing the local reaction rate with the volume average, we inherently assume that concentration fields in all the spheres within an electrode are identical. Intuitively, this is not a very liming assumption when computing the cell voltage. Because, the cell voltage as derived in Eq. 5.31 has surface concentrations at the two ends of the cell, the point where the current enters or leaves the cell. Finally, the cell voltage can be rewritten as follows:

$$\begin{aligned}V_{\text{cell}} &= V_{0,p}\left(c_{sur,p}\right)|_{x=L} - V_{0,n}\left(c_{sur,n}\right)|_{x=0} + \phi_{l,x=L} - \phi_{l,x=0}\\ &- \frac{2R_gT}{F}\sinh^{-1}\left(\frac{I}{2j_{p0}\left(c_{sur,p}\right)a_pL_pF}\right)|_{x=L} - \frac{2R_gT}{F}\sinh^{-1}\left(\frac{I}{2j_{n0}\left(c_{sur,n}\right)a_nL_nF}\right)|_{x=0}\end{aligned} \tag{5.48}$$

The above equation still has one limitation that nothing is known about the electrolyte potentials. But before we reach there, let us see some fascinating insights of the model developed thus far.

5.3.4 *Insights from the Reduced Order Model*

During discharge, in the negative electrode, when a neutral solid phase Li is released into the electrolyte as a Li^+ ion and an electron is released into the outer circuit. This electron recombines with some other Li^+ in electrolyte within the positive electrode region (almost instantaneously, neglecting electronic transients), and forms a solid phase Li. Thus, if a solid phase Li decreases in the negative electrode, a solid phase Li gets formed at the positive electrode. Thus, an overall solid phase lithium balance holds at any time.

$$L_n\varepsilon_{sn}\langle \bar{c}_{sn}\rangle + L_p\varepsilon_{sp}\langle \bar{c}_{sp}\rangle = L_n\varepsilon_{sn}c_{sn0} + L_p\varepsilon_{sp}c_{sp0} = \text{Constant.} \tag{5.49}$$

This equation shows that the average solid phase lithium concentration in the two electrodes are not independent of each other. For example, if the average solid phase concentration in the negative electrode is known, which in the positive electrode is computed from Eq. 5.49 as

$$\langle \bar{c}_{sp} \rangle = c_{sp0} + \frac{L_n \varepsilon_{sn}}{L_p \varepsilon_{sp}} \left(c_{sn0} - \langle \bar{c}_{sn} \rangle \right) \tag{5.50}$$

This equation is very useful in computations, as concentration of one electrode can be written in terms of the other electrode. Intuitively, only electrons move in and out of a sealed lithium cell. Hence, an overall lithium balance is obviously maintained. Due to electro-neutrality at any time the total amounts of solid phase Li and hence, electrolyte phase Li^+ ions are individually constant. Hence, there are separate Li balances for the solid (Eq. 5.49) and electrolyte phases. As noted earlier, these volume averaged results are exact, even though they entail loss of concentration profile or gradient information.

At this point, let us dwell deep into the singular shortcoming of the reduced order model, lack of information about the potential drop in the electrode and the electrolyte.

5.3.5 Electrode Potential Drop

For obtaining the potentials, using the uniform reaction rate approximation, the local reaction rate is assumed to be the volume average reaction rate, at any time, i.e.

$$j_n (x, t) \approx \langle j_n \rangle (t) = \frac{I}{a_n L_n F} \tag{5.51}$$

and for the positive electrode:

$$j_p (x, t) \approx \langle j_p \rangle (t) = -\frac{I}{a_p L_p F} \tag{5.52}$$

With the uniform reaction rate approximation, the negative electrode solid current balance Eq. 5.33 becomes

$$\frac{\partial}{\partial x} \left(-\sigma_{sn} \frac{\partial \phi_s}{\partial x} \right) \approx -a_n F \langle j_n \rangle (t) = -\frac{I(t)}{L_n} \tag{5.53}$$

The above equation is integrated with respect to x to get the following:

$$-\sigma_{sn} \frac{\partial \phi_s}{\partial x} = -\frac{I(t)}{L_n} x + f(t) \tag{5.54}$$

where $f(t)$ is some arbitrary function. Using the boundary condition Eq. 5.34 gives

$$0 = -\frac{I(t)}{L_n} L_n + f(t) \tag{5.55}$$

Eliminating $f(t)$ in Eq. 5.54 using Eq. 5.55 gives the solid current in negative electrode, under uniform reaction rate approximation, as follows:

$$-\sigma_{sn}\frac{\partial \phi_s}{\partial x} = \frac{I(t)}{L_n}(L_n - x) \tag{5.56}$$

Note that Eq. 5.56 satisfies the other boundary condition Eq. 5.55. Similarly, the solid current in the positive electrode, under uniform reaction approximation is as follows:

$$-\sigma_{sp}\frac{\partial \phi_s}{\partial x} = \frac{I(t)}{L_p}\left[x - (L_n + L_s)\right] \tag{5.57}$$

Thus uniform reaction rate approximation implies a linear solid current profile in the electrodes, i.e., the solid current density linearly varies from the total current density at the collector end to zero at the separator end. At this point, let us revisit our reduced order model and account for the voltage drop due to electronic conduction.

5.3.6 Voltage Due to Electronic Conduction

Typically, the electronic conductivity of electrode is much larger than the electrolyte leading to minimal voltage loss. However, electrode materials with low conductivity—typically of the positive electrode materials—leads to an Ohmic drop across the positive electrode determined as follows:

$$V_{ec} = \int_{L_n+L_s}^{L} \frac{d\phi_{sp}}{dx} dx. \tag{5.58}$$

Using Eq. (5.57) in the above equation,

$$\begin{aligned} V_{ec} &= \int_{L_n+L_s}^{L} -\frac{I}{\sigma_{sp} L_p}\left(x - (L_n + L_s)\right) dx, \\ &= -\frac{I L_p}{2\sigma_{sp}}. \end{aligned} \tag{5.59}$$

The Ohmic potential drop is therefore added to the cell voltage as follows:

$$V = \phi_{sp}|_{x=L} - \phi_{sn}|_{x=0} - \frac{I L_p}{2\sigma_{sp}} \tag{5.60}$$

This results in the final form for the cell voltage form the reduced order model to be as follows:

$$V_{\text{cell}} = V_{0,p}\left(c_{sur,p}\right)|_{x=L} - V_{0,n}\left(c_{sur,n}\right)|_{x=0} + \phi_{l,x=L} - \phi_{l,x=0}$$
$$- \frac{2R_gT}{F}\sinh^{-1}\left(\frac{I}{2j_{p0}\left(c_{sur,p}\right)a_pL_pF}\right)|_{x=L} - \frac{2R_gT}{F}\sinh^{-1}\left(\frac{I}{2j_{n0}\left(c_{sur,n}\right)a_nL_nF}\right)|_{x=0} - \frac{IL_p}{2\sigma_{sp}} \tag{5.61}$$

Finally, we have accounted for all the process in a normal cell operation in the reduced order model, except the electrolyte potential. Accounting this, by far, will be most evolved due to following reasons:

- Lack of any symmetry in the equations unlike the solid phase mass balance
- Mass and charge transfer fields are intricately coupled, at every point in the domain. Hence consistent solution of both the equations is required.
- Electrolyte phase being continuous across both the electrode and the separator regions. Constructing a solution accounting the boundary conditions and the continuity conditions, would result in coupling of the variables.

5.4 Treatment of the Electrolyte Potential

5.4.1 Simple Linear and Nonlinear Approaches

A simplest approach would be to assume a linear resistance, $\phi_{l,x=L} - \phi_{l,x=0} = IR_{el}$ that can be added to the reduced order model. In line with the electronic conduction, we can infer that $R_{el} \sim L/\kappa_l$. This will make the model equation look like the following:

$$V_{\text{cell}} = V_{0,p}\left(c_{sur,p}\right)|_{x=L} - V_{0,n}\left(c_{sur,n}\right)|_{x=0}$$
$$- \frac{2R_gT}{F}\sinh^{-1}\left(\frac{I}{2j_{p0}\left(c_{sur,p}\right)a_pL_pF}\right)|_{x=L} - \frac{2R_gT}{F}\sinh^{-1}\left(\frac{I}{2j_{n0}\left(c_{sur,n}\right)a_nL_nF}\right)|_{x=0}$$
$$- I\left[\frac{L_p}{2\sigma_{sp}} + \frac{L_p}{\kappa_{lp}} + \frac{L_s}{\kappa_{ls}} + \frac{L_n}{\kappa_{ln}}\right] \tag{5.62}$$

A nonlinear version of the electrolyte can be developed by noticing that larger voltage transients must, intuitively, have larger electrolyte potential drops. This information is absent in the definition of a constant electrolyte resistance model used in the earlier equation. Thus, $R_{el} \sim \frac{L}{\kappa_l}V_{cell}^{\gamma}$, where γ accounts for the nonlinear response of the electrolyte field. Thus the reduced order model is, eventually the following:

$$V_{\text{cell}} = V_{0,p}\left(c_{sur,p}\right)|_{x=L} - V_{0,n}\left(c_{sur,n}\right)|_{x=0}$$
$$-\frac{2R_gT}{F}\sinh^{-1}\left(\frac{I}{2j_{p0}\left(c_{sur,p}\right)a_pL_pF}\right)|_{x=L} - \frac{2R_gT}{F}\sinh^{-1}\left(\frac{I}{2j_{n0}\left(c_{sur,n}\right)a_nL_nF}\right)|_{x=0}$$
$$-I\left[\frac{L_p}{2\sigma_{sp}} + \frac{L_p}{\kappa_{lp}}V_{cell}^{\gamma_p} + \frac{L_s}{\kappa_{ls}}V_{cell}^{\gamma_s} + \frac{L_n}{\kappa_{ln}}V_{cell}^{\gamma_n}\right] \tag{5.63}$$

However, γ does not have any physical significance, and we are deviating from our basic promise of a fully physically relevant model framework. Hence, let us revert back to our approach of volume averaging and profile approximations.

5.4.2 *Volume Averaging the Electrolyte Mass Balance Equation*

The first step toward obtaining the electrolyte potential is to look at the electrolyte mass balance equation for the two electrode sections and the separator. As this equation is first order in time and second order in space. As attempted previously, the approach is as follows:

- The volume averaging results in the time evolution equation for the electrolyte concentrations. This results in appearance of the flux at the interfaces as a relevant variable
- Invoke profile approximations for the concentrations, resulting in concentration profile in terms of the flux at the interfaces, and concentrations.
- Decouple the set of equations, such that each variable is obtained by solving a single equation

The diffusion of Li ions form the electrolyte phase mass balances. Li ion enters the electrolyte field through the Butler–Volmer reaction any one electrode, and leaves through the other electrode. Hence the reaction form source or sink terms in the electrode regions, while electrolyte accumulates and diffuses through the separator. Electrolyte concentration field is continuous through the three regions: negative electrode, separator, and positive electrode. Hence, the concentration and flux are continuous at the electrode—separator interfaces. Let us define interfacial concentrations (c_{2in}, c_{2ip}) and fluxes defined by (q_{2in}, q_{2ip}) are identified as separate variables, so that it is easy to impose them in the individual regions, and maintain their equality across interfaces. At the collector ends zero flux boundary conditions apply.

5.4.2.1 Negative Electrode Region

The mass balance of the electrolyte in the negative electrode region is as follows:

$$\varepsilon_{ln}\frac{\partial c_l}{\partial t} = \frac{\partial}{\partial x}D_{ln}\frac{\partial c_l}{\partial x} + a_n(1-t_+)j_n \tag{5.64}$$

with boundary conditions

$$\begin{aligned} -D_{ln}\frac{\partial c_l}{\partial x}|_{x=0} &= 0 \\ c_l(L_n, t) &= c_{lin} \\ -D_{ln}\frac{\partial c_l}{\partial x}|_{x=L_n} &= q_{lin} \end{aligned} \tag{5.65}$$

The electrolyte diffusion equation in the negative electrode region, Eq. 5.64 is volume averaged as follows:

$$\frac{1}{L_n}\int_{x=0}^{L_n}\left[\varepsilon_{ln}\frac{\partial c_l}{\partial t} = \frac{\partial}{\partial x}\left(D_{ln}\frac{\partial c_l}{\partial x}\right) + a_n(1-t_+)j_n\right]dx \tag{5.66}$$

The volume average of the time derivative becomes the time derivative of the volume average.

$$LHS = \frac{\varepsilon_{ln}}{L_n}\int_{x=0}^{L_n}\frac{\partial c_l}{\partial t}dx = \varepsilon_{ln}\frac{d}{dt}\left(\frac{1}{L_n}\int_{x=0}^{L_n}c_l dx\right) = \varepsilon_{ln}\frac{d\langle c_l\rangle_n}{dt} \tag{5.67}$$

The diffusive flux gradient term in the right hand side of Eq. 5.64 gets volume averaged, using the boundary conditions Eq. 5.65 as follows:

$$\begin{aligned} \langle\frac{\partial}{\partial x}\left(D_{ln}\frac{\partial c_n}{\partial x}\right)\rangle_n &= \frac{1}{L_n}\int_{x=0}^{L_n}\frac{\partial}{\partial x}\left(D_{ln}\frac{\partial c_n}{\partial x}\right)dx \\ &= \frac{1}{L_n}\left(D_{ln}\frac{\partial c_n}{\partial x}\right)_{x=0}^{L_n} = -\frac{q_{lin}}{L_n} \end{aligned} \tag{5.68}$$

The volume average of the reaction rate term in the right hand side of Eq. 5.64 is straight forward as follows:

$$\langle a_n(1-t_+)j_n\rangle_n = \frac{1}{L_n}\int_{x=0}^{L_n}a_n(1-t_+)j_n dx = a_n(1-t_+)\langle j_n\rangle \tag{5.69}$$

Putting together these results and multiplying by the electrode thickness, the volume averaged electrolyte diffusion equation in the negative electrode region is as follows:

$$L_n \varepsilon_{ln} \frac{d \langle c_l \rangle_n}{dt} = -q_{lin} + (1 - t_+) L_n a_n \langle j_n \rangle \tag{5.70}$$

While deriving the volume-averaged equation for the negative electrode region, the volume average of the diffusive flux gradient term in the electrolyte diffusion equation is given by the exact result, Eq. 5.68. To approximate the concentration profile in the negative electrode region, it is assumed that the following:

$$\frac{\partial}{\partial x}\left(D_{nl} \frac{\partial c_n}{\partial x}\right) \approx -\frac{q_{lin}}{L_n} \tag{5.71}$$

This expression is the lowest order approximation which on volume averaging will satisfy the exact result, Eq. 5.68. The profile approximation Eq. 5.71 is integrated once with respect to x as follows:

$$\frac{\partial}{\partial x}\left(D_{nl} \frac{\partial c_n}{\partial x}\right) = -\frac{q_{lin}}{L_n} = D_{nl} \frac{\partial c_n}{\partial x} = -\frac{q_{lin}}{L_n} x + f(t) \tag{5.72}$$

where $f(t)$ is some arbitrary function. Applying the boundary condition at separator interface Eq. 5.65 gives

$$x = L_n : f(t) = 0; \; D_{nl} \frac{\partial c_n}{\partial x} = -\frac{q_{lin}}{L_n} x \tag{5.73}$$

Note that Eq. 5.73 satisfies the collector end boundary condition, Eq. 5.65. Rearranging Eq. 5.73 and integrating once more with respect to x gives

$$\frac{\partial c_l}{\partial x} = -\frac{q_{lin}}{L_n D_{ln}} x; \; c_l(x,t) = -\frac{q_{lin}}{L_n D_{ln}} \frac{x^2}{2} + f(t) \tag{5.74}$$

where f(t) is some arbitrary function. Applying the interfacial concentration boundary condition of Eq. 5.65 gives

$$x = L_n; \; c_{lin} = -\frac{q_{lin}}{L_n D_{ln}} \frac{L_n^2}{2} + f(t) \tag{5.75}$$

Note that the interfacial concentration appears as a natural variable. Hence, it is also considered as a physically relevant internal variable along with the interfacial flux. Eliminating $f(t)$ between Eqs. 5.74 and 5.75 gives

$$c_l(x,t) = c_{lin}(t) + \frac{q_{lin}(t)}{2 L_n D_{ln}} \left(L_n^2 - x^2\right) \tag{5.76}$$

Once the interfacial concentration and flux are known, the electrolyte concentration profile is computed using Eq. 5.76. The rest of the model development focuses on deriving equations for the interfacial concentration and flux. Applying the volume averaging definition for the quadratic term appearing in Eq. 5.76 gives

$$\langle c_l \rangle_n = c_{lin} + \frac{L_n q_{lin}}{3D_{ln}} \tag{5.77}$$

Differentiating this equation with respect to time gives

$$\frac{d\langle c_l \rangle_n}{dt} = \frac{dc_{lin}}{dt} + \frac{L_n q_{lin}}{3D_{ln}} \frac{dq_{lin}}{dt} \tag{5.78}$$

Using the above expression in Eq. 5.70 gives

$$L_n \varepsilon_{ln} \frac{dc_{lin}}{dt} + \frac{L_n \varepsilon_{lin}}{3D_{ln}} \frac{dq_{lin}}{dt} = -q_{lin} + (1 - t_+)\, a_n L_n \langle j_n \rangle \tag{5.79}$$

Equation 5.79 is one equation for the two unknowns: interfacial concentration and interfacial flux in the negative electrode region. One more equation is needed for these two unknowns for closure. These variables are related to the interfacial concentration and flux in the positive electrode region, due to conservation of the electrolyte within the cell. The derivation, however, can be extended to accommodate concentration dependent electrolyte diffusivity. For later reference, from Eq. 5.76 the electrolyte concentration near the negative collector end is obtained as follows:

$$c_l\,(x = 0, t) = c_{lin}\,(t) + \frac{q_{lin}\,(t)\, L_n}{2D_{ln}} \tag{5.80}$$

5.4.2.2 Positive Electrode Region

$$\varepsilon_{lp} \frac{\partial c_l}{\partial t} = \frac{\partial}{\partial x} D_{lp} \frac{\partial c_l}{\partial x} + a_p (1 - t_+) j_p \tag{5.81}$$

with boundary conditions

$$-D_{lp} \frac{\partial c_l}{\partial x}|_{x=L_n+L_s} = q_{lip};\ c_l(L_n + L_s, t) = c_{lip};\ -D_{lp} \frac{\partial c_l}{\partial x}|_{x=L} = 0 \tag{5.82}$$

Across the three regions, the initial condition is as follows:

$$c_l(x, 0) = c_{l0} \tag{5.83}$$

In the positive electrode region, the diffusive flux term in the right hand side of Eq. 5.81 gets volume averaged, using the boundary conditions Eq. 5.82 as follows:

$$\langle\frac{\partial}{\partial x}\left(D_{lp}\frac{\partial c_n}{\partial x}\right)\rangle_p = \frac{1}{L_p}\int_{x=L_n+L_s}^{L}\frac{\partial}{\partial x}\left(D_{lp}\frac{\partial c_p}{\partial x}\right)dx = \frac{1}{L_n}\left(D_{lp}\frac{\partial c_p}{\partial x}\right)_{x=L_n+L_s}^{L} = \frac{q_{lip}}{L_p} \tag{5.84}$$

Volume averaging of the accumulation and reaction rate terms in Eq. 5.81 is as before. Finally, the volume averaged electrolyte diffusion equation in the positive electrode region is as follows:

$$L_p\varepsilon_{lp}\frac{\langle c_n\rangle_p}{dt} = q_{lip} + (1-t_+)\,a_pL_p\langle j_p\rangle \tag{5.85}$$

The derivation is analogous to the one above, so only the salient steps are presented. From Eq. 5.85, the profile approximation is derived as follows:

$$\frac{\partial}{\partial x}\left(D_{lp}\frac{\partial c_l}{\partial x}\right) \approx \frac{q_{lip}}{L_p} \tag{5.86}$$

This approximation is integrated with respect to x once and the boundary condition at the collector end Eq. 5.82 is applied to get the following:

$$D_{lp}\frac{\partial c_l}{\partial x} = -\frac{q_{lip}}{L_p}(L-x) \tag{5.87}$$

Note that this equation satisfies the interfacial flux boundary condition Eq. 5.82. Integrating once again with respect to x and applying the interfacial concentration boundary condition Eq. 5.82 leads to the following:

$$c_l(x,t) = c_{lip}(t) - \frac{q_{lip}(t)}{2L_pD_{lp}}\left[L_p^2 - (L-x)^2\right] \tag{5.88}$$

Using the volume average definition and the volume averaging result

$$\langle c_l\rangle_p = c_{lip} - \frac{L_pq_{lip}}{3D_{lp}} \tag{5.89}$$

Differentiating this with respect to time and using it in Eq. 5.85 gives

$$L_p\varepsilon_{lp}\frac{dc_{lip}}{dt} - \frac{L_p^2\varepsilon_{2p}}{3D_{lp}}\frac{dq_{lip}}{dt} = q_{lip} + (1-t_+)\,a_pL_p\langle j_p\rangle \tag{5.90}$$

For later, from Eq. 5.88 the electrolyte concentration near the positive collector end is obtained as follows:

$$c_l\left(x = L, t\right) = c_{lip}\left(t\right) - \frac{q_{lip}\left(t\right) L_p}{2D_{lp}} \tag{5.91}$$

5.4.2.3 Separator Region

$$\varepsilon_{ls}\frac{\partial c_l}{\partial t} = \frac{\partial}{\partial x} D_{ls}\frac{\partial c_l}{\partial x} \tag{5.92}$$

with boundary conditions

$$-D_{ln}\frac{\partial c_l}{\partial x}|_{x=0} = q_{lin};\ c_l(L_n, t) = c_{lin};\ -D_{ln}\frac{\partial c_l}{\partial x}|_{x=L_n+L_s} = q_{lip};\ c_l(L_n + L_s, t) = c_{lip}; \tag{5.93}$$

The volume average definitions for a function f(x,t) in the negative and positive electrode regions are as defined earlier. The corresponding definition in the separator region is as follows:

$$\langle f\rangle_s(t) = \frac{1}{AL_s}\int_{x=L_n}^{x=L_n+L_s} f(x,t)Adx = \frac{1}{L_s}\int_{x=L_n}^{L_n+L_s} f(x,t)dx \tag{5.94}$$

In the separator region, the diffusive flux gradient term in the right hand side of Eq. 5.92 gets volume averaged, using the boundary conditions Eq. 5.93 as follows:

$$\langle\frac{\partial}{\partial x}\left(D_{ln}\frac{\partial c_n}{\partial x}\right)\rangle_s = \frac{1}{L_s}\int_{x=L_n}^{L_n+L_s}\frac{\partial}{\partial x}\left(D_{ns}\frac{\partial c_l}{\partial x}\right)dx = \frac{1}{L_s}\left(D_{ns}\frac{\partial c_l}{\partial x}\right)_{x=L_n}^{x=L_n+L_s} = \frac{q_{lin}}{L_s} - \frac{q_{lip}}{L_s} \tag{5.95}$$

Volume averaging of the accumulation term in Eq. 5.92 is as before. Finally, the volume averaged electrolyte diffusion equation in the separator region is as follows:

$$L_s\varepsilon_{ls}\frac{\langle c_n\rangle_s}{dt} = q_{lin} - q_{lip} \tag{5.96}$$

From Eq. 5.95 the profile approximation is derived as follows:

$$\frac{\partial}{\partial x}\left(D_{ls}\frac{\partial c_l}{\partial x}\right) \approx \frac{q_{lin} - q_{lip}}{L_s} \tag{5.97}$$

The electrolyte diffusion equation, Eq. 5.92 shows that the above expression is the lowest order approximation which can support an accumulation of electrolyte in the separator region. Electrolyte accumulation or depletion in the separator is expected during high current transients. This expression is integrated once with respect to x, and the negative electrode side flux boundary condition Eq. 5.93 is applied to give:

$$D_{ls}\frac{\partial c_l}{\partial x} = -q_{lin} + \frac{q_{lin} - q_{lip}}{L_s}(x - L_n) \tag{5.98}$$

Note that this equation satisfies the positive electrode side flux boundary condition. Integrating once again with respect to x and applying the negative electrode side interfacial concentration boundary condition Eq. 5.93 gives the following:

$$c_l(x,t) = c_{lin}(t) - \frac{q_{lin}(t)}{D_{ls}}(x - L_n) + \left(\frac{q_{lin}(t) - q_{lip}(t)}{L_s D_{ls}}\right)\frac{(x - L_n)^2}{2} \tag{5.99}$$

Note that the positive electrode side interfacial concentration boundary condition Eq. 5.93 is not yet utilized. Applying it on Eq. 5.99 gives an algebraic equation between the interfacial concentrations and fluxes

$$c_{lin} - c_{lip} = \frac{L_s\left(q_{lin} + q_{lip}\right)}{2D_{ls}}; \tag{5.100}$$

This is the first equation relating the interface concentrations and fluxes.

Using the definition of volume averaging, the average concentration in the separator region is given by

$$\langle c_l \rangle_s = c_{lin} - \frac{L_s q_{lin}}{3D_{ls}} - \frac{L_s q_{lip}}{6D_{ls}} \tag{5.101}$$

Thus, the profile approximations in the three regions have yielded three equations: Eqs. 5.79, 5.90 and 5.100 for the four unknowns: two interfacial concentrations and two interfacial fluxes. The fourth equation is stipulated among these four unknowns by the overall electrolyte balance.

5.4.2.4 Overall Electrolyte Balance

The evolution equations for the average electrolyte concentrations in the three regions, Eqs. 5.70, 5.96 and 5.85, are not independent of each other. During discharge, in the negative electrode, when an Li^+ ion gets into the electrolyte an electron gets released into the outer circuit. This electron recombines with some other Li^+ in the electrolyte within the positive electrode region. Thus, if a Li^+ ion is released in the negative electrode region, some other Li^+ ion gets consumed in the positive electrode region. Hence, an overall electrolyte balance holds at any time

$$L_n\varepsilon_{ln}\langle c_l\rangle_n + L_s\varepsilon_{ls}\langle c_l\rangle_s + L_p\varepsilon_{lp}\langle c_l\rangle_p = \left(L_n\varepsilon_{ln} + L_s\varepsilon_{ls} + L_p\varepsilon_{lp}\right)c_{l0} \tag{5.102}$$

where c_{l0} is the initial electrolyte concentration, as the initial condition. Differentiating this equation with respect to time gives

$$L_n \varepsilon_{ln} \frac{d\langle c_l \rangle_n}{dt} + L_s \varepsilon_{ls} \frac{d\langle c_l \rangle_s}{dt} + L_p \varepsilon_{lp} \frac{d\langle c_p \rangle_n}{dt} = 0 \tag{5.103}$$

Substituting the average concentrations of the three regions Eqs. 5.77, 5.101, and 5.89 in the overall electrolyte balance Eq. 5.103, and rearranging gives

$$(L_n \varepsilon_{ln} + L_s \varepsilon_{ls}) c_{lin} + \left(\frac{L_n^2 \varepsilon_{ln}}{3D_{ln}} - \frac{L_s^2 \varepsilon_{ls}}{3D_{ls}} \right) q_{lin} + L_p \varepsilon_{lp} c_{lip} - \left(\frac{L_p^2 \varepsilon_{lp}}{3D_{lp}} + \frac{L_s^2 \varepsilon_{ls}}{6D_{ls}} \right) q_{lip} = (L_n \varepsilon_{ln} + L_s \varepsilon_{ls} + L_p \varepsilon_{lp}) c_{l0} \tag{5.104}$$

Now, with four equations for the four unknowns, the formulation is closed. For later reference, from Eq. 5.99 the electrolyte concentration in the middle of the separator is obtained as

$$c_{lmid} = c_l \left(L_n + \frac{L_s}{2}, t \right) = c_{lin} - \frac{3L_s}{8D_{ls}} q_{lin} - \frac{L_s}{8D_{ls}} q_{lip} \tag{5.105}$$

In the volume averaged electrolyte diffusion equations discussed above, the interfacial fluxes appear explicitly, hence they are identified as physically relevant internal variables characterizing the cell behavior. As noted earlier, these volume averaged results are exact and not approximations, even though they entail loss of concentration profile or gradient information.

5.4.2.5 Decoupling the Equations

Using the volume averaging, the complete electrolyte domain information is contained in 4 variables—two intermediate concentrations, c_{lin}, c_{lip} and the two fluxes q_{lin}, q_{lip}. The four corresponding equations are Eqs. 5.79, 5.90, 5.100 and 5.104, and it can be seen that these equations are couped and cannot be used directly. In this subsection, we will decouple and derive one equation for each variable of interest. Let us rearrange Eq. 5.100 to write c_{lin} in terms of c_{lip}.

$$c_{lin} = c_{lip} + \frac{L_s \left(q_{lin} + q_{lip} \right)}{2D_{ls}}; \tag{5.106}$$

Using this in Eq. 5.104 gives

$$c_{lip} = c_{l0} + \alpha_{in} q_{lin} + \alpha_{ip} q_{lip} \tag{5.107}$$

where the new constants are defined by

$$\alpha_n = -\frac{\frac{L_n L_s \varepsilon_n}{2D_{ls}} + \frac{L_s^2 \varepsilon_s}{6D_{ls}} + \frac{L_n^2 \varepsilon_n}{3D_{ln}}}{L_n \varepsilon_n + L_s \varepsilon_s + L_p \varepsilon_p} \tag{5.108}$$

and

$$\alpha_p = -\frac{\frac{L_p L_s \varepsilon_p}{2D_{ls}} + \frac{L_s^2 \varepsilon_s}{6D_{ls}} + \frac{L_p^2 \varepsilon_p}{3D_{ln}}}{L_n \varepsilon_n + L_s \varepsilon_s + L_p \varepsilon_p} \tag{5.109}$$

Differentiating Eq. 5.107 with respect to time, we get the derivative of the interface concentration:

$$\frac{dc_{lip}}{dt} = \alpha_{in}\frac{dq_{lin}}{dt} + \alpha_{ip}\frac{dq_{lip}}{dt} \tag{5.110}$$

Differentiating Eq. 5.104 with respect to time and using Eq. 5.110, we get an equivalent relation for c_{lin}.

$$\frac{dc_{lin}}{dt} = \left(\alpha_{in} + \frac{L_s}{2D_{ls}}\right)\frac{dq_{lin}}{dt} + \left(\alpha_{ip} + \frac{L_s}{2D_{ls}}\right)\frac{dq_{lip}}{dt} \tag{5.111}$$

The two equations above relate the interfacial concentration derivatives to interfacial flux derivatives. This can be used in Eqs. 5.79 and 5.90 to eliminate the former variables resulting in:

$$\left(L_n\varepsilon_{ln}\alpha_{in} + \frac{L_s L_n \varepsilon_{ln}}{2D_{2s}} + \frac{L_n^2 \varepsilon_{2n}}{3D_{2n}}\right)\frac{dq_{lin}}{dt} + \left(L_n\varepsilon_{ln}\alpha_{ip} + \frac{L_n L_s \varepsilon_{ln}}{2D_{2s}}\right)\frac{dq_{lip}}{dt}$$
$$= -q_{lin} + (1 - t_+)\, a_n L_n \langle j_n \rangle \tag{5.112}$$

and

$$\left(L_p\varepsilon_{lp}\alpha_{in}\right)\frac{dq_{lin}}{dt} + \left(L_p\varepsilon_{lp}\alpha_{ip} - \frac{l_p^2 \varepsilon_{lp}}{3D_{lp}}\right)\frac{dq_{lip}}{dt} = q_{lip} + (1 - t_+)\, a_p L_p \langle j_p \rangle \tag{5.113}$$

Although with this step, we have eliminated the interface concentrations, still the interface fluxes are coupled. We re-arrange the equations a bit to result, eventually in:

$$\mathscr{D}\frac{dq_{lin}}{dt} = -B_2 q_{lin} - A_2 q_{lip} + A_3 B_2 \langle j_n \rangle - A_2 B_3 \langle j_p \rangle \tag{5.114}$$

and

$$\mathscr{D}\frac{dq_{lip}}{dt} = B_1 q_{lin} + A_1 q_{lip} - A_3 B_1 \langle j_n \rangle + A_1 B_3 \langle j_p \rangle \tag{5.115}$$

for the interface flux in the negative and the positive electrode domains respectively. The coupling constants in the above equations are given by

$$
\begin{aligned}
A_1 &= L_n \varepsilon_{ln} \alpha_{in} + \frac{L_s L_n \varepsilon_{ln}}{2D_{2s}} + \frac{l_n^2 \varepsilon_{2n}}{3D_{2n}} \\
A_2 &= L_p \varepsilon_{lp} \alpha_{ip} - \frac{l_p^2 \varepsilon_{lp}}{3D_{lp}} \\
A_3 &= (1 - t_+)\, a_n L_n \\
B_1 &= L_p \varepsilon_{lp} \alpha_{in} \\
B_2 &= L_p \varepsilon_{lp} \alpha_{ip} - \frac{l_p^2 \varepsilon_{lp}}{3D_{lp}} \\
B_3 &= (1 - t_+)\, a_p L_p \\
\mathscr{D} &= A_1 B_2 - A_2 B1
\end{aligned}
\tag{5.116}
$$

In this subsection, we have derived the electrolyte concentration field. The outcome of volume averaging is to identify that the interface fluxes and concentrations are the salient variables. The time evolution of the interface flux alone needs to be tracked, and analytical expressions are developed to obtain the interfacial concentrations. These concentrations, in turn are used to obtain the concentration at the ends of the cell - the relevant variables required for the cell potential.

5.5 Electrolyte Potential Field

5.5.1 *Field Equations*

The total current at any point in the cell is the sum of solid current (Ohmic component in the solid phase),migration current (electrolyte migration due to electric potential gradient) and diffusion current (current carried by diffusive flux caused by concentration gradient). In the separator only migration and diffusion currents exist. The electrolyte potential field is continuous across the three regions; hence, like electrolyte concentration field, electrolyte potential and migration current continuities apply at the electrode-separator interfaces.At the collector ends, migration currents go to zero and solid currents equal the external current. At the electrode-separator interfaces, solid currents go to zero, so that the sum of migration and diffusion currents equals the total current. The interfacial electrolyte potential and interfacial migration currents are identified as separate variables, so that is easy to impose them in the individual regions, and maintain their equality across the interfaces.

5.5.1.1 Separator Region

$$- \kappa_{ls} \frac{\partial \phi_l}{\partial x} + 2\kappa_{ls} \frac{R_g T}{F} (1 - t_+) \frac{\partial \ln c_l}{\partial x} = I \tag{5.117}$$

where

$$\theta = \frac{R_g T}{F} (1 - t_+) \tag{5.118}$$

with boundary conditions

$$\begin{aligned} -\kappa_{\ln} \frac{\partial \phi_l}{\partial x} |_{x=L_n} &= I_{lin}(t) \\ \phi_l(L_n, t) &= \phi_{lin}(t) \\ -\kappa_{\ln} \frac{\partial \phi_l}{\partial x} |_{x=L_n+L_s} &= I_{lip}(t) \\ \phi_l(L_n + L_s, t) &= \phi_{lip}(t) \end{aligned} \tag{5.119}$$

One needs to assume a reference potential (either for electrolyte or solid potential) somewhere within the cell. This work takes a zero reference potential at the center of the separator

$$\phi_l \left(L_n + \frac{L_s}{2}, t \right) = 0 \tag{5.120}$$

Having an electrolyte potential reference within the separator eases out analytical integration in the reduced order model development as shown below, since solid current component is absent in the separator total current balance Eq. 5.117. The cell voltage is the difference in solid potential between the two collector ends. Hence, the choice of reference potential (both location and value) does not affect the cell voltage.

The electrolyte concentration profile in the separator is known from earlier section and using this profile, Eq. 5.117 can be directly integrated to get the electrolyte potential profile in the separator. Ionic conductivity is often a strong function of electrolyte concentration. In the reduced order framework, however, it is laborious to handle spatial variation of material properties. Due to the thinness of the electrode or separator regions, ionic conductivity is assumed to be a function of the volume average concentration, instead of the local concentration:

$$\kappa_{ls}(x,t) = \kappa_l [c_l(x,t)] = \kappa_l [\langle c_l \rangle_s (x,t)] \tag{5.121}$$

Note that the temporal variations of the ionic conductivity are accounted through average electrolyte concentration. Equation 5.117 is rearranged to get

$$\frac{\partial \phi_l}{\partial x} = 2\theta \frac{\partial ln c_l}{\partial x} - \frac{I(t)}{\kappa_{ls}(t)} \tag{5.122}$$

Integrating, and applying the datum:

$$\phi_l(x,t) = 2\theta ln\left[\frac{c_l(x,t)}{c_{l,mid}(x,t)}\right] - \frac{I(t)}{\kappa_{ls}(t)}\left[x - \left(L_n + \frac{L_s}{s}\right)\right] \tag{5.123}$$

This is an exact result, subject to the diffusion approximations in the electrolyte concentration field and the approximation in Eq. 5.121. Using the above expression the electrolyte potential at the two interfaces is identified as

$$\phi_{lin}(x,t) = \phi_l(L_n,t) = 2\theta ln\left[\frac{c_{lin}(x,t)}{c_{l,mid}(x,t)}\right] + \frac{I(t)\,I_s}{2\kappa_{ls}(t)} \tag{5.124}$$

$$\phi_{lip}(x,t) = \phi_l(L_n + L_s,t) = 2\theta ln\left[\frac{c_{lip}(x,t)}{c_{l,mid}(x,t)}\right] - \frac{I(t)\,I_s}{2\kappa_{ls}(t)} \tag{5.125}$$

These interfacial electrolyte potentials will be used while solving for the electrolyte potential in the electrode regions, to ensure electrolyte potential continuity.

5.5.1.2 Negative Electrode Region

$$-\sigma_{sn}\frac{\partial\phi_s}{\partial x} - \kappa_{ln}\frac{\partial\phi_l}{\partial x} + 2\kappa_{ln}\frac{R_g T}{F}(1 - t_+)\frac{\partial ln c_l}{\partial x} = I \tag{5.126}$$

with boundary conditions

$$\begin{aligned} -\kappa_{ln}\frac{\partial\phi_l}{\partial x}|_{x=0} &= 0 \\ \phi_l(L_n,t) &= \phi_{lin}(t) \\ -\kappa_{ln}\frac{\partial\phi_l}{\partial x}|_{x=L_n} &= I_{lin}(t) \end{aligned} \tag{5.127}$$

As before the ionic conductivity is assumed to be a function of the average electrolyte concentration in this region

$$\kappa_{ls}(x,t) = \kappa_l[c_l(x,t)] = \kappa_l[\langle c_l\rangle_s(x,t)] \tag{5.128}$$

The solid current term appears in the total current balance for negative electrode (Eq. 5.126). The expression for solid current is taken from Eq. 5.56, rewritten in a once integrated form to facilitate integration of the total current balance equation,

$$-\sigma_{sn}\frac{\partial\phi_s}{\partial x} = \frac{I(t)}{L_n}(L_n - x) = \frac{\partial}{\partial x}\left[-\frac{I(t)}{L_n}\frac{(L_n - x)^2}{2}\right] \tag{5.129}$$

Note that the uniform reaction rate approximation is entering into electrolyte potential calculation through this solid current equation. Using Eq. 5.129 in the total current balance Eq. 5.126 gives

$$\frac{\partial}{\partial x}\left[-\frac{I(t)}{L_n}\frac{(L_n - x)^2}{2}\right] - \kappa_{ln}\frac{\partial \phi_l}{\partial x} + 2\kappa_{ln}\theta\frac{\partial lnc_l}{\partial x} = I \tag{5.130}$$

This equation can be rearranged into a total differential as

$$\frac{\partial}{\partial x}\left[-\frac{I(t)}{L_n}\frac{(L_n - x)^2}{2} - \kappa_{ln}\phi_l + 2\kappa_{ln}\theta lnc_l\right] = I \tag{5.131}$$

Ionic conductivity, according to equation Eq. 5.128, does not have spatial dependence. Hence, the above equation is easily put in total differential form. Integrating once with respect to x, and applying the boundary condition of total current gives

$$\phi_l(x,t) = \phi_{lin}(t) + 2\theta ln\left(\frac{c_l(x,t)}{c_{lin}(x,t)}\right) + \frac{I(t)}{\kappa_{ln}}(L_n - x) - \frac{I(t)}{2\kappa_{ln}}(L_n - x)^2 \tag{5.132}$$

This is an exact result, subject to the diffusion approximations in the electrolyte concentration field and uniform reaction rate approximation in the solid current expression. For later reference, from Eq. 5.132 the electrolyte potential near the negative collector end is obtained as

$$\phi_l(x=0,t) = \phi_{lin}(t) + 2\theta ln\left(\frac{c_l(x=0,t)}{c_{lin}(x,t)}\right) + \frac{I(t)L_n}{2\kappa_{ln}} \tag{5.133}$$

5.5.1.3 Positive Electrode Section

$$-\sigma_{sp}\frac{\partial \phi_s}{\partial x} - \kappa_{lp}\frac{\partial \phi_l}{\partial x} + 2\kappa_{lp}\frac{R_g T}{F}(1 - t_+)\frac{\partial lnc_l}{\partial x} = I \tag{5.134}$$

with boundary conditions

$$\begin{aligned} -\kappa_{lp}\frac{\partial \phi_l}{\partial x}|_{x=L} &= 0 \\ \phi_l(L_n + L_s, t) &= \phi_{lip}(t) \\ -\kappa_{lp}\frac{\partial \phi_l}{\partial x}|_{x=L_n+L_s} &= I_{lip}(t) \end{aligned} \tag{5.135}$$

The solution procedure is similar to the one above. The solid current equation is integrated and interfacial potential boundary condition is used:

$$\phi_l(x,t) = \phi_{lip}(t) + 2\theta ln\left(\frac{c_l(x,t)}{c_{lip}(x,t)}\right) - \frac{I(t)}{\kappa_{lp}}[x-(L_n+L_s)] + \frac{I(t)}{2\kappa_{lp}L_p}[x-(L_n+L_s)]^2 \tag{5.136}$$

For later reference, from Eq. 5.136 the electrolyte potential near the positive collector end is obtained as

$$\phi_l(x=L,t) = \phi_{lip}(t) + 2\theta ln\left(\frac{c_l(x=L,t)}{c_{lip}(x,t)}\right) - \frac{I(t)L_p}{2\kappa_{lp}} \tag{5.137}$$

With Eqs. 5.132, 5.136 and 5.123 the electrolyte potential field is fully solved.

5.5.2 Revisiting the Battery Model

In this section, we derived the equation for the electrolyte potential, as a function of the interface concentrations and fluxes. Once we know the potential at two ends of the cell, the difference gives the voltage drop due to electrolyte. The step wise procedure to obtain this quantity is enumerated below. For ease of usage, the corresponding equations are also written with the steps. Hence the reader can skip the detailed derivations and use the equations directly for the required application.

1. Obtain the uniform reaction rate

$$\begin{aligned}\langle j_n\rangle &= \frac{I}{a_nL_nF}\\ \langle j_p\rangle &= -\frac{I}{a_pL_pF}\end{aligned} \tag{5.138}$$

2. Obtain α_{in}, α_{ip} and other constants:

$$\alpha_n = -\frac{\frac{L_nL_s\varepsilon_n}{2D_{ls}} + \frac{L_s^2\varepsilon_s}{6D_{ls}} + \frac{L_n^2\varepsilon_n}{3D_{ln}}}{L_n\varepsilon_n + L_s\varepsilon_s + L_p\varepsilon_p}$$

$$\alpha_p = -\frac{\frac{L_pL_s\varepsilon_p}{2D_{ls}} + \frac{L_s^2\varepsilon_s}{6D_{ls}} + \frac{L_p^2\varepsilon_p}{3D_{ln}}}{L_n\varepsilon_n + L_s\varepsilon_s + L_p\varepsilon_p}$$

$$A_1 = L_n\varepsilon_{ln}\alpha_{in} + \frac{L_sL_n\varepsilon_{ln}}{2D_{2s}} + \frac{l_n^2\varepsilon_{2n}}{3D_{2n}}$$

$$
\begin{aligned}
A_2 &= L_p \varepsilon_{lp} \alpha_{ip} - \frac{l_p^2 \varepsilon_{lp}}{3 D_{lp}} \\
A_3 &= (1 - t_+) \, a_n L_n \\
B_1 &= L_p \varepsilon_{lp} \alpha_{in} \\
B_2 &= L_p \varepsilon_{lp} \alpha_{ip} - \frac{l_p^2 \varepsilon_{lp}}{3 D_{lp}} \\
B_3 &= (1 - t_+) \, a_p L_p \\
\mathscr{D} &= A_1 B_2 - A_2 B1
\end{aligned}
\tag{5.139}
$$

3. Solve the ODE for interfacial flux at the positive and negative electrode region

$$
\begin{aligned}
\mathscr{D} \frac{dq_{lin}}{dt} &= -B_2 q_{lin} - A_2 q_{lip} + A_3 B_2 \langle j_n \rangle - A_2 B_3 \langle j_p \rangle \\
\mathscr{D} \frac{dq_{lip}}{dt} &= B_1 q_{lin} + A_1 q_{lip} - A_3 B_1 \langle j_n \rangle + A_1 B_3 \langle j_p \rangle
\end{aligned}
\tag{5.140}
$$

4. Solve for the electrolyte concentrations at the (a) electrode—electrolyte interfaces, (b) middle of separator (datum), and at (c) ends of the cell

$$
\begin{aligned}
c_{lin} &= c_{lip} + \frac{L_s \left(q_{lin} + q_{lip} \right)}{2 D_{ls}} \\
c_{lip} &= c_{l0} + \alpha_{in} q_{lin} + \alpha_{ip} q_{lip} \\
c_{lmid} &= c_{lin} - \frac{3 L_s}{8 D_{ls}} q_{lin} - \frac{L_s}{8 D_{ls}} q_{lip} \\
c_l \left(x = 0, t \right) &= c_{lin} \left(t \right) + \frac{q_{lin} \left(t \right) L_n}{2 D_{ln}} \\
c_l \left(x = L, t \right) &= c_{lip} \left(t \right) - \frac{q_{lip} \left(t \right) L_p}{2 D_{lp}}
\end{aligned}
\tag{5.141}
$$

5. Solve for the potentials at the two ends of the cell, and hence the voltage drop as the difference between the two:

$$
\begin{aligned}
\phi_l \left(x = 0, t \right) &= \phi_{lin} \left(t \right) + 2\theta ln \left(\frac{c_l \left(x = 0, t \right)}{c_{lin} \left(x, t \right)} \right) + \frac{I \left(t \right) L_n}{2 \kappa_{ln}} \\
\phi_l \left(x = L, t \right) &= \phi_{lip} \left(t \right) + 2\theta ln \left(\frac{c_l \left(x = L, t \right)}{c_{lip} \left(x, t \right)} \right) - \frac{I \left(t \right) L_p}{2 \kappa_{lp}}
\end{aligned}
\tag{5.142}
$$

6. Obtain the voltage due to electrolyte filed:

$$
V_{el} = \phi_l \left(x = L, t \right) - \phi_l \left(x = 0, t \right) \tag{5.143}
$$

7. Finally, obtain the cell voltage that includes the electrolyte voltage drop (Table 5.2):

$$V_{\text{cell}} = V_{el} + V_{0,p}\left(c_{sur,p}\right)|_{x=L} - V_{0,n}\left(c_{sur,n}\right)|_{x=0} \\ - \frac{2R_gT}{F}\sinh^{-1}\left(\frac{I}{2j_{p0}\left(c_{sur,p}\right)a_pL_pF}\right)|_{x=L} - \frac{2R_gT}{F}\sinh^{-1}\left(\frac{I}{2j_{n0}\left(c_{sur,n}\right)a_nL_nF}\right)|_{x=0} \tag{5.144}$$

Table 5.2 Table gives a compilation of governing equations for the reduced order model

Negative Electrode	Positive Electrode
Solid phase mass balance	
$\frac{d\langle\bar{c}_{sn}\rangle}{dt} = -\frac{3\langle j_n\rangle}{r_n}$	$\frac{d\langle\bar{c}_{sp}\rangle}{dt} = -\frac{3\langle j_p\rangle}{r_p}$
$\frac{d\langle\bar{c}_{srn}\rangle}{dt} = -\frac{30D_{sn}\langle\bar{c}_{srn}\rangle}{r_n^2} - \frac{45\langle j_n\rangle}{2r_n^2}$	$\frac{d\langle\bar{c}_{srp}\rangle}{dt} = -\frac{30D_{sp}\langle\bar{c}_{srp}\rangle}{r_p^2} - \frac{45\langle j_p\rangle}{2r_p^2}$
$c_{sur,n} = \langle\bar{c}_{sn}\rangle + \frac{8r_n}{35}\langle\bar{c}_{srn}\rangle - \frac{r_n}{35D_{sn}j_n}$	$c_{sur,p} = \langle\bar{c}_{sp}\rangle + \frac{8r_p}{35}\langle\bar{c}_{srp}\rangle - \frac{r_p}{35D_{sp}j_p}$
Charge transfer reaction	
$j_n = \frac{I}{a_nL_nF}$	$j_p = -\frac{I}{a_pL_pF}$
Electrolyte Potential	
$A_1 = \left(L_n\varepsilon_{ln}\alpha_{in} + \frac{L_sL_n\varepsilon_{ln}}{2D_{2s}} + \frac{l_n^2\varepsilon_{2n}}{3D_{2n}}\right)$; $A_2 = \left(L_n\varepsilon_{ln}\alpha_{ip} + \frac{L_nL_s\varepsilon_{ln}}{2D_{2s}}\right)$; $A_3 = (1-t_+)\,a_nL_n$	
$B_1 = \left(L_p\varepsilon_{lp}\alpha_{in}\right)$; $B_2 = \left(L_p\varepsilon_{lp}\alpha_{ip} - \frac{l_p^2\varepsilon_{lp}}{3D_{lp}}\right)$; $B_3 = (1-t_+)\,a_pL_p$; $\mathscr{D} = A_1B_2 - A_2B_1$.	
$c_{lmid} = c_{lin} - \frac{3L_sq_{lin}}{8D_{ls}} - \frac{L_sq_{lip}}{8D_{ls}}$	
$\alpha_n = -\frac{\frac{L_nL_s\varepsilon_n}{2D_{ls}} + \frac{L_s^2\varepsilon_s}{6D_{ls}} + \frac{L_n^2\varepsilon_n}{3D_{ln}}}{L_n\varepsilon_n + L_s\varepsilon_s + L_p\varepsilon_p}$	$\alpha_p = -\frac{\frac{L_pL_s\varepsilon_p}{2D_{ls}} + \frac{L_s^2\varepsilon_s}{6D_{ls}} + \frac{L_p^2\varepsilon_p}{3D_{ln}}}{L_n\varepsilon_n + L_s\varepsilon_s + L_p\varepsilon_p}$
$c_{lin} = c_{lip} + \frac{L_s(q_{lin}+q_{lip})}{2D_{ls}}$	$c_{lip} = c_{l0} + \alpha_nq_{lin} + \alpha_pq_{lip}$
$c_{ln} = c_{lin} + \frac{q_{lin}L_n}{2D_{ln}}$	$c_{lp} = c_{lip} - \frac{q_{lip}L_p}{2D_{lp}}$
$\frac{dq_{lin}}{dt} =$ $-\frac{B_2}{\mathscr{D}}q_{lin} - \frac{A_2}{\mathscr{D}}q_{lip} + \frac{A_3B_2}{\mathscr{D}}\langle j_n\rangle - \frac{A_2B_3}{\mathscr{D}}\langle j_p\rangle$	$\frac{dq_{lip}}{dt} =$ $\frac{B_1}{\mathscr{D}}q_{lin} + \frac{A_1}{\mathscr{D}}q_{lip} - \frac{A_3B_1}{\mathscr{D}}\langle j_n\rangle + \frac{A_1B_3}{\mathscr{D}}\langle j_p\rangle$
$\phi_{lin} = 2\theta ln\frac{c_{lin}}{c_{lmid}} + \frac{IL_s}{2\kappa_{ls}}$	$\phi_{lip} = 2\theta ln\frac{c_{lip}}{c_{lmid}} - \frac{IL_sep}{2\kappa_{ls}}$
$V_{el,n} = \phi_{lin} + 2\theta ln\frac{c_{ln}}{c_{lin}} + \frac{IL_n}{2\kappa_{ln}}$	$V_{el,p} = \phi_{lip} + 2\theta ln\frac{c_{lp}}{c_{lip}} - \frac{IL_p}{2\kappa_{lp}}$
$V_{rd,n} =$ $\frac{2R_gT}{F}\sinh^{-1}\left(\frac{j_n}{2k_nc_{sur,n}^0.5(c_{smax,n}-c_{sur,n})^0.5c_{ln}^0.5)}\right)$	$V_{rd,p} =$ $\frac{2R_gT}{F}\sinh^{-1}\left(\frac{j_p}{2k_pc_{sur,p}^0.5(c_{smax,p}-c_{sur,p})^0.5c_{ln}^0.5)}\right)$
Cell Voltage	
Cell OCV	$U_{cell} = U_p - U_n$
Electrolyte voltage drop	$V_{el} = V_{el,p} - V_{el,n}$
Reaction diffusion voltage drop	$V_{rd} = V_{rd,p} - V_{rd,n}$
$V_{cell} = U_{cell} + V_{el} + V_{rd}$	

5.6 Developments to the Base ROM

The ROM presented in this chapter, in addition to reducing the PDEs to simple analytical expressions and ODEs, provides a generic approach using which further developments can be done. As we graduated from EC model to ECT in the previous chapters, we will add the thermal balance in the first subsection. Inclusion of degradation and cycling, thereby providing an estimate of the battery cycle life will be considered next. The intriguing effects of phase change materials will be considered as the last subsection.

5.6.1 Thermal Balance

The thermal balance with local contribution to heat is given by equation were developed in the last chapters. A recap is provided here:

$$\frac{\partial\left(\rho_k C_{pk} T_k\right)}{\partial t} = \frac{\partial}{\partial x}\left(\lambda_k \frac{\partial T_k}{\partial x}\right) + Q, \tag{5.145}$$

$$\text{With BCs,} \quad -\lambda_k \frac{\partial T_k}{\partial x}|_{x=0,L} = h_k (T_k - T_0), \quad -\lambda_k \frac{\partial T_k}{\partial x}|_{x=L_n, L_n+L_s} = Q_{ik}. \tag{5.146}$$

where ρ, C_p and h denote the density, specific heat capacity and the heat transfer coefficient of the system component while Q_{ik} denotes the interfacial heat flux. Q is the total heat generated with contributions from various heat sources, namely, the irreversible heat from the overpotential (Q_{irrk}), reversible heat due to the system entropy (Q_{revk}) and the Ohmic heat from the electrode (Q_{Ohmsk}) and electrolyte phases (Q_{Ohmlk}).

$$Q = Q_{rev_k} + Q_{irr_k} + Q_{Ohm_{sk}} + Q_{Ohm_{lk}}, \tag{5.147}$$

$$Q_{rev_k} = F a_k j_k T_k \frac{\partial U_k}{\partial T}, \tag{5.148}$$

$$Q_{irr_k} = F a_k j_k \left(\phi_{sk} - \phi_{lk} - U_k\right), \tag{5.149}$$

$$Q_{Ohm_{sk}} = \sigma_k \left(\frac{\partial \phi_{sk}}{\partial x}\right)^2, \tag{5.150}$$

$$Q_{Ohm_{lk}} = \kappa_k \left(\frac{\partial \phi_{lk}}{\partial x}\right)^2 + \kappa_{Dk} \frac{1}{c_{lk}} \frac{\partial c_{lk}}{\partial x} \frac{\partial \phi_{lk}}{\partial x}. \tag{5.151}$$

where $\kappa_{Dk} = -2\kappa R_g T_k (1 - t_+)/F \times (1 + d\ln f_\pm / d\ln c_{lk})$. In order to use the definition of volume average as discussed earlier on each of the terms in Eq. (5.145), we tackle the accumulation term:

$$\langle \frac{\partial\left(\rho_k C_{pk} T_k\right)}{\partial t} \rangle = \rho_k C_{pk} \frac{d\langle T_k \rangle}{dt} \tag{5.152}$$

The first time on the right hand side of equation, the conduction term Eq. (5.145) and the BCs (Eq. 5.146) are used to obtain,

$$\langle\frac{\partial}{\partial x}\left(\lambda_k\frac{\partial T_k}{\partial x}\right)\rangle = \frac{\mp Q_{in,p} \pm h_k\left(T_k|_{x=0,L} - T_0\right)}{L_k}. \tag{5.153}$$

Although the steps were direct, when we look at the energy source terms (5.148–5.151) it can be noted that the terms are not linear and using the definition of volume averaging involves sophisticated integration techniques. Volume averaging of coupled terms are very involved. To obtain closed solutions, we would need to assume that the volume average of the product of two terms is equal to the product of averages. Therefore, to simplify volume average of Q_{rev_k}, Eqs. 5.39 and 5.40 are made use of,

$$\langle Q_{rev_k}\rangle = \frac{I}{L_k}\langle T_k\rangle\frac{dU_k}{dT}. \tag{5.154}$$

In the case of Q_{irr_k}, the value of ϕ_{sk} is substituted from Eq. 5.30 and then subjected to volume averaging, to obtain,

$$\langle Q_{irr_k}\rangle = \frac{I}{L_k}\left[\frac{2R\langle T_k\rangle}{F}\sinh^{-1}\left(\frac{j_k}{j_{k0}}\right)\right]. \tag{5.155}$$

The definition of Ohmic heats involves product of potential gradients, the values of which are substituted from Eqs. 5.56 and 5.57. Averaging these over the electrode volumes give,

$$\langle Q_{Ohm_{sk}}\rangle = \frac{I^2}{3\sigma_{sk}}, \tag{5.156}$$

$$\langle Q_{Ohm_{lk}}\rangle = \frac{I^2}{3\kappa_{lk}} + \frac{2I\theta q_{lik}}{3c_{lk}D_{lk}}. \tag{5.157}$$

Using Eqs. 5.152–5.157, the thermal balance equations averaged over the electrode and separator volumes, are obtained as,

$$\rho_n C_{pn}\frac{d\langle T_n\rangle}{dt} = \frac{-Q_{in} + h_n(T_n - T_0)}{L_n} + \frac{I}{L_n}\left[\frac{2RT}{F}\sinh^{-1}\left(\frac{j_n}{j_{n0}}\right)\right] \tag{5.158}$$
$$+ \frac{I}{L_n}\langle T_n\rangle\frac{dU_n}{dT} + \frac{I^2}{3\sigma_n} + \frac{2I\theta q_{2in}}{3c_{2n}D_{2n}} + \frac{I^2}{3\kappa_n},$$

$$\rho_p C_{pp}\frac{d\langle T_p\rangle}{dt} = \frac{Q_{ip} - h_p(T_p - T_0)}{L_p} - \frac{I}{L_p}\left[\frac{2RT}{F}\sinh^{-1}\left(\frac{j_p}{j_{p0}}\right)\right] \tag{5.159}$$

$$+\frac{I}{L_p}\langle T_p\rangle\frac{dU_p}{dT} + \frac{I^2}{3\sigma_p} + \frac{2I\theta q_{2ip}}{3c_{2p}D_{2p}} + \frac{I^2}{3\kappa_p},$$

$$\rho_s C_{ps}\frac{d\langle T_s\rangle}{dt} = \frac{Q_{in} - Q_{ip}}{L_s} - \frac{I\left(\phi_{2s}|_{x=L_n+L_s} - \phi_{2s}|_{x=L_n}\right)}{L_s}. \tag{5.160}$$

To obtain the temperature profile within the electrodes and separator, the local gradient of temperature is equated to the average value. That is,

$$\frac{\partial}{\partial x}\left(\lambda_k\frac{\partial T_k}{\partial x}\right) = \langle\frac{\partial}{\partial x}\left(\lambda_k\frac{\partial T_k}{\partial x}\right)\rangle, \tag{5.161}$$

$$= \frac{\mp Q_{in,p} \pm h_k\left(T_k|_{x=0,L} - T_0\right)}{L_k}. \tag{5.162}$$

Integrating the Eq. 5.162 and using the boundary conditions (Eq. 5.146) the temperature within the electrodes is obtained as,

$$T_n = T_{in} + \frac{-Q_{in}}{2\lambda_n L_n}\left(x^2 - L_n^2\right) + \frac{h_n\left(T_n|_{x=0} - T_0\right)}{2\lambda_n L_n}\left(x - L_n\right)^2, \tag{5.163}$$

$$T_p = T_{ip} + \frac{Q_{ip}}{2\lambda_p L_p}\left(x^2 - (L_n + L_s)^2 - 2L(x - (L_n + L_s))\right) \tag{5.164}$$

$$-\frac{h_p\left(T_p|_{x=L} - T_0\right)}{2\lambda_p L_p}\left(x^2 - (L_n + L_s)^2\right)^2,$$

$$T_s = T_{in} + \frac{Q_{in}}{2\lambda_s L_s}\left(x^2 - L_n^2 - 2(L_n + L_s)(x - L_n)\right) - \frac{Q_{ip}}{2\lambda_s ls}\left(x^2 - L_n^2\right)^2. \tag{5.165}$$

where T_{ik} is the temperature at the electrode-separator interface. It may be noted that the four unknown quantities namely, the interfacial temperatures (T_{ik}) and heat fluxes (Q_{ik}) are involved in the Eqs. (5.163–5.165). To obtain a solution for these variables, the temperature profiles are volume averaged to obtain,

$$\langle T_n\rangle = T_{in} + \frac{Q_{in}L_n}{3\lambda_n} + \frac{h_n\left(T_n|_{x=0} - T_0\right)L_n}{6\lambda_n}, \tag{5.166}$$

$$\langle T_p\rangle = T_{ip} - \frac{Q_{ip}L_p}{3\lambda_p} - \frac{h_p\left(T_p|_{x=L} - T_0\right)L_p}{6\lambda_p}, \tag{5.167}$$

$$\langle T_s\rangle = T_{in} - \frac{Q_{in}L_s}{3\lambda_s} - \frac{Q_{ip}L_s}{6\lambda_s}. \tag{5.168}$$

Differentiating these equations with respect to time and substituting the values of average temperatures in Eqs. 5.158–5.160, respectively, differential equations in terms of the interfacial variables are obtained.

$$\rho_n C_{pn}\left(\frac{dT_{in}}{dt}+\frac{L_n}{3\lambda_n}\frac{dQ_{in}}{dt}\right)=\frac{-Q_{in}+h_n\left(T_n|_{x=0}-T_0\right)}{L_n} \tag{5.169}$$

$$+\frac{I}{L_n}\left[\frac{2RT}{F}\sinh^{-1}\left(\frac{j_n}{j_{n0}}\right)\right]+\frac{I}{L_n}\langle T_n\rangle\frac{dU_n}{dT}+\frac{I^2}{3\sigma_n}+\frac{2I\theta q_{2in}}{3c_{2n}D_{2n}}+\frac{I^2}{3\kappa_n},$$

$$\rho_p C_{pp}\left(\frac{dT_{ip}}{dt}-\frac{L_p}{3\lambda_p}\frac{dQ_{ip}}{dt}\right)=\frac{Q_{ip}-h_p\left(T_p|_{x=L}-T_0\right)}{L_p} \tag{5.170}$$

$$-\frac{I}{L_p}\left[\frac{2RT}{F}\sinh^{-1}\left(\frac{j_p}{j_{p0}}\right)\right]+\frac{I}{L_p}\langle T_p\rangle\frac{dU_p}{dT}+\frac{I^2}{3\sigma_p}+\frac{2I\theta q_{2ip}}{3c_{2p}D_{2p}}+\frac{I^2}{3\kappa_p},$$

$$\rho_s C_{ps}\left(\frac{dT_{in}}{dt}-\frac{L_s}{2\lambda_s}\frac{dQ_{in}}{dt}-\frac{L_s}{6\lambda_s}\frac{dQ_{ip}}{dt}\right)=\frac{Q_{in}-Q_{ip}}{L_s} \tag{5.171}$$

$$-\frac{I\left(\phi_{2s}|_{x=L_n+L_s}-\phi_{2s}|_{x=L_n}\right)}{L_s}.$$

Using the condition $T_s|_{x=L_n+L_s}=T_{ip}$ in the Eq. 5.165 leads to fourth equation in terms of the interfacial variables.

$$T_{in}-T_{ip}=\frac{L_s(Q_{in}+Q_{ip})}{2\lambda_s}. \tag{5.172}$$

Solving the Eqs. 5.169–5.172, the four unknown interfacial variables (T_{in}, T_{ip}, Q_{in}) and (Q_{ip}) are determined. Knowing the interfacial variables the average temperatures are calculated using Eqs. 5.166, 5.167 and 5.168.

The temperatures at the current collector ends are obtained from Eqs. 5.163 and 5.164 to be,

$$T_n|_{x=0}=\left[T_{in}+\frac{Q_{in}L_n}{2\lambda_n}-\frac{h_nT_0L_n}{2\lambda_n}\right]/\left[1-\frac{h_nL_n}{2\lambda_n}\right], \tag{5.173}$$

$$T_p|_{x=L}=\left[T_{ip}-\frac{Q_{ip}L_p}{2\lambda_p}+\frac{h_pT_0L_p}{2\lambda_p}\right]/\left[1+\frac{h_pL_p}{2\lambda_p}\right]. \tag{5.174}$$

5.6.2 Coupling of Electrochemical and Thermal Models

The cell properties namely, the diffusion coefficients, conductivities and the reaction rate constants depend on the cell temperature which in turn depends on the properties and cell voltage. To take the dependencies into account, the model is solved by a two way coupling. A flowchart depicting the two way coupling is given in Fig. 5.1.

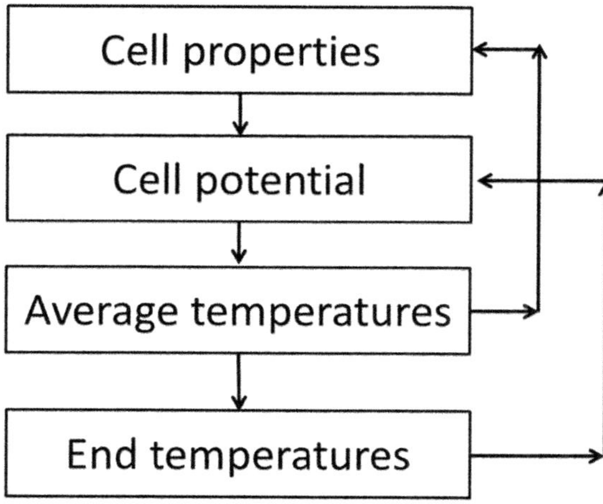

Fig. 5.1 A flowchart showing the two way coupling of the model

The cell potentials, calculated using Eq. 5.30, are used to calculate the temperature (Eqs. 5.166–5.168). The temperature dependent electrolyte phase conductivity and diffusion coefficient are obtained from fits to experiments, while the temperature dependence of solid phase diffusion coefficient and reaction rate constant is introduced in the model through an Arrhenius equation. The average values of temperature (Eqs. 5.166–5.168) are used to obtain the modified values of cell transport properties. The temperatures at the current collector ends (Eqs. 5.173 and 5.174) are used to obtain the solid potentials (5.30) at those points which are used to obtain the cell voltage.

5.6.3 *Degradation: Cycling Studies*

In the present work, the side reaction involving electrolyte reduction at the negative electrode is considered to be the main source of degradation. The total current at the negative electrode is split into the component responsible for the lithium intercalation/deintercalation reaction and the one responsible for the side reaction. This is explained using a schematic in Fig. 5.2.

$$\Rightarrow j_n(x,t) = j_{ni}(x,t) + j_s(x,t) \tag{5.175}$$

The side reaction pertains to loss of lithium at the interface of the negative electrode to form SEI. The side reaction rate can be modeled using Tafel kinetics [9] as,

$$j_s(x,t) = \frac{-j_{s0}}{F}\exp\left(-\frac{\alpha_c F}{R_g T}\left(\phi_{1n} - \phi_{2n} - U_s - j_n(x,t)R_f F\right)\right). \tag{5.176}$$

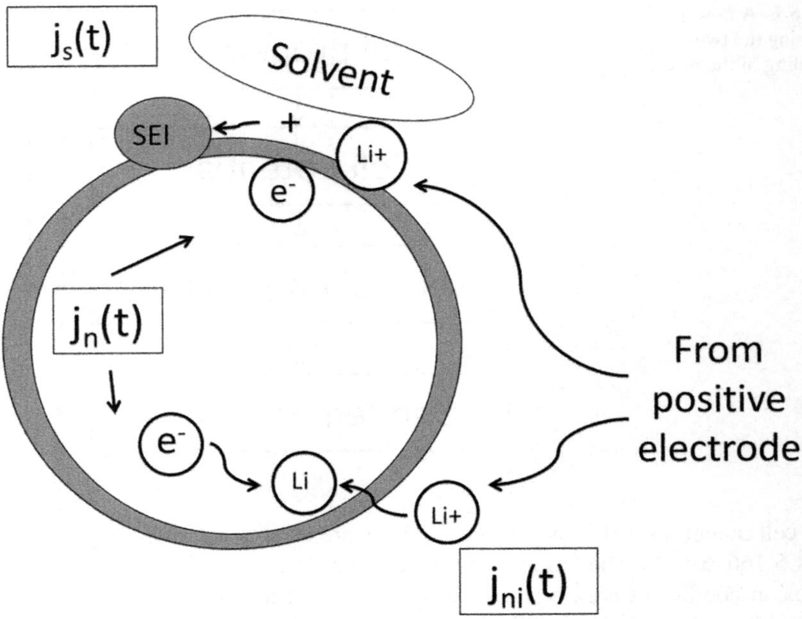

Fig. 5.2 A schematic representation of the SEI formation on an active material particle surface

The side reaction results in the formation of a resistive film of reaction products on the electrode surface (SEI). It should be noted that the overpotential driving the side reaction is affected by the Ohmic drop across this film. The film resistance is calculated as,

$$R_f = R_{f_0} + \frac{\delta_f}{\kappa_f}, \tag{5.177}$$

where, κ_f is the conductivity of the SEI film while δ_f is the film thickness which is obtained from,

$$\frac{d\delta_f}{dt} = -\frac{j_s M_f}{\rho_f}. \tag{5.178}$$

That is, the film evolution depends on its molecular weight (M_f), density (ρ_f) and the rate of the side reaction. To take into account the degradation, the Eq. 5.30 for the negative electrode is modified to,

$$\phi_{sn}|_{x=0} = \phi_{2n}|_{x=0} + U_n(c_{sn}) + j_n|_{x=0} F R_f|_{x=0} + \frac{2R_g T}{F}\sinh^{-1}\left(\frac{j_n|_{x=0} - j_s|_{x=0}}{2j_{n0}}\right) \tag{5.179}$$

In the absence of side reaction at the positive electrode, Eq. (5.30) remains unchanged except for the intercalation reaction current ($j_{pi}(x,t)$) being replaced with the total current ($j_p(x,t)$), that is,

$$\phi_{sp}|_{x=L} = \phi_{lp}|_{x=L} + U_p(c_{sp}) + \frac{2R_gT}{F}\sinh^{-1}\left(\frac{j_p|_{x=L}}{2j_{p0}}\right). \tag{5.180}$$

To determine the cell potential using Eqs. (5.179) and (5.180), the unknown variables are the surface concentration of lithium in the solid phase and the potentials in the electrolyte phase at the current collectors. The procedure to obtain these variables is outlined next. For the degradation because of SEI growth, there is no volume averaging required, as the pertinent equation is an ODE.

5.6.4 Phase Change

In this subsection, we consider an intriguing feature of some positive electrodes that undergo a concentration dependent phase change during the insertion/deinsertion of lithium. The key step towards modeling a phase change system is to describe the solid phase diffusion accurately. During a major portion of the charge/discharge cycle of the cell, multiphases can coexist. The diffusion within the active material is expected to be affected due to this coexistence. Traditionally modeling using an isotropic shrinking core approach, recent experimental studies reveal direction in the crystal plane and the phase boundary propagates anisotropically throughout the particle. To incorporate phase change, a concentration-dependent diffusion coefficient developed assuming the system to be a mixture of two phases.

In this regard, the arguments used in the coupling of the pseudo 2D model are applied to this case. In the pseudo 2D model for a porous electrode, it is assumed that the electrode is a homogeneous mixture of both the solid and electrolyte phases. This enables the model to couple the solid phase concentration, obtained by solving the solid mass balance equations (in radial coordinates), to the electrolyte equations, at any point along the thickness of the cell. In a phase change system, irrespective of the mode of the phase boundary propagation, the active material particle consists of the two coexisting phases. A schematic of the arguments presented here is provided in the Fig. 5.3. From the schematic it can be seen that, along the radius of the particle, one encounters a mixture of both the phases.

Accordingly, the diffusivity of the phase change active material particle can be expressed as,

$$D_{sp_{\text{mix}}} = x_{p\alpha}D_{sp\alpha} + x_{p\beta}D_{sp\beta} \tag{5.181}$$

where, $x_{p\alpha}$, $x_{p\beta}$ are the mole fractions of the lithium-deficient phase, α and the lithium rich phase, β, respectively and $D_{sp\alpha}$, $D_{sp\beta}$ are their diffusion coefficients. $x_{sp\alpha}$ and $x_{sp\beta}$ are given by the following equations,

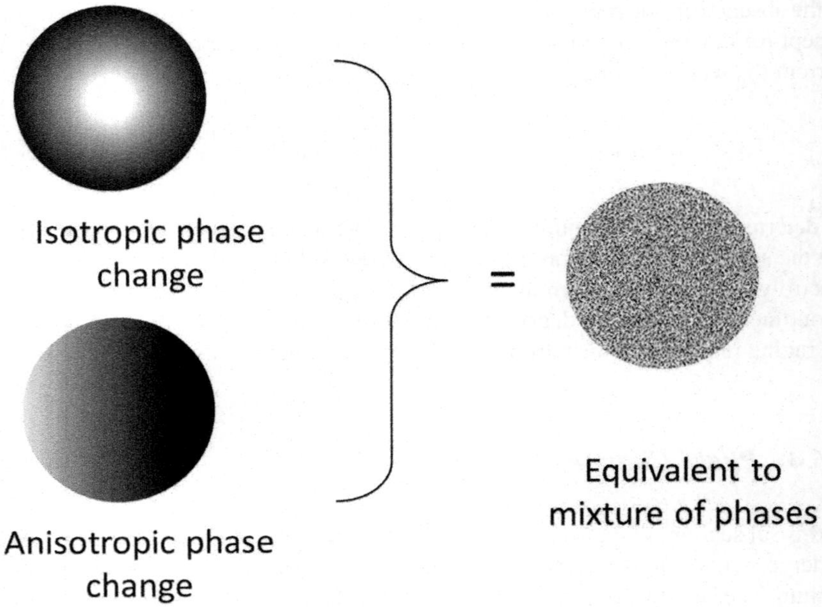

Fig. 5.3 A schematic representation of the isotropic shrinking core approach and the anisotropic phase propagation approach adopted for a phase change system in literature. Irrespective of the model of phase formation, in the present work, the system is considered to be equivalent to a mixture of two phases

$$x_{sp\alpha} = \frac{c_{sp\beta} - \langle c_{sp} \rangle}{c_{sp\beta} - c_{sp\alpha}} \tag{5.182}$$

$$x_{sp\beta} = 1 - x_{sp\alpha} \tag{5.183}$$

It can be noted that the diffusivity varies with the average concentration, c_{sp} while taking into account the equilibrium concentrations and diffusivities of the individual phases (α and β) present in the system. When c_{sp} is equal to either $c_{sp\alpha}$ or $c_{sp\beta}$, the diffusion coefficient is equal to the pure phase diffusivity, $D_{s\alpha}$ and $D_{s\beta}$, respectively.

The solid phase mass balance for the positive electrode undergoing phase change can be modified by the new definition of diffusivity:

$$\frac{\partial c_{sp}}{\partial t} = \frac{1}{r^2}\frac{\partial}{\partial r}\left(r^2 D_{sp_{\text{mix}}}\frac{\partial c_{sp}}{\partial r}\right) \tag{5.184}$$

$$\text{With the BCs } -D_{sp_{\text{mix}}}\frac{\partial c_{sp}}{\partial r}|_{r=R_p} = j_{pi},\ -D_{sp_{\text{mix}}}\frac{\partial c_{sp}}{\partial r}|_{r=0} = 0 \tag{5.185}$$

The negative electrode equations remains without any change.

5.7 Summary of the ROM

In this section, we will summarize how to solve the ROM model to obtain the cell voltage. The reader could cut across the derivations and use the steps described here as a cookbook. Hence, in this section, we have focused on the solution sequence and not the physical significance. For the latter, the reader would refer to the relevant sections in the chapter. The steps described below gives the cell voltage including the degradation due to SEI, at a constant temperature. In order to use a non-isothermal version, one would need to additionally solve Eqs. 5.173 and 5.174, for the temperatures at the ends of cell.

It can be seen that the complete battery model, that involved solution of 10 coupled PDEs can be reduced to a set of 6 ODEs, and a set of simple analytical expressions. The analytical expressions and the ODEs are completely decoupled and hence the solution non-iterative. The complete framework is achieved without loss of any physical insight—all the variables retain their respective physical significance. This can be considered as the most prominent achievement, in comparison with many other state of art ROM methods. In the next chapter, we will see the multitude of applications the ROM can be used accurately. In addition, the ROM enables easy integration with estimators and can be used onboard for state and health estimation.

1. Constants to be solved for the following:

$$
\begin{aligned}
\alpha_n &= -\frac{\frac{L_n L_s \varepsilon_n}{2D_{ls}} + \frac{L_s^2 \varepsilon_s}{6D_{ls}} + \frac{L_n^2 \varepsilon_n}{3D_{ln}}}{L_n \varepsilon_n + L_s \varepsilon_s + L_p \varepsilon_p} \\
\alpha_p &= -\frac{\frac{L_p L_s \varepsilon_p}{2D_{ls}} + \frac{L_s^2 \varepsilon_s}{6D_{ls}} + \frac{L_p^2 \varepsilon_p}{3D_{ln}}}{L_n \varepsilon_n + L_s \varepsilon_s + L_p \varepsilon_p} \\
A_1 &= L_n \varepsilon_{ln} \alpha_{in} + \frac{L_s L_n \varepsilon_{ln}}{2D_{2s}} + \frac{l_n^2 \varepsilon_{2n}}{3D_{2n}} \\
A_2 &= L_p \varepsilon_{lp} \alpha_{ip} - \frac{l_p^2 \varepsilon_{lp}}{3D_{lp}} \\
A_3 &= (1 - t_+)\, a_n L_n \\
B_1 &= L_p \varepsilon_{lp} \alpha_{in} \\
B_2 &= L_p \varepsilon_{lp} \alpha_{ip} - \frac{l_p^2 \varepsilon_{lp}}{3D_{lp}} \\
B_3 &= (1 - t_+)\, a_p L_p \\
\mathscr{D} &= A_1 B_2 - A_2 B1
\end{aligned}
\tag{5.186}
$$

2. ODEs to be solved for:

$$\frac{d\langle\bar{c}_{sn}\rangle}{dt} = -3\frac{I}{a_n L_n F r_n} \tag{5.187}$$

$$\frac{d\langle\bar{c}_{sp}\rangle}{dt} = 3\frac{I}{a_p L_p F r_p} \tag{5.188}$$

$$\frac{d\langle\bar{c}_{snr}\rangle}{dt} + \frac{30 D_{sn}}{r_k^2}\bar{c}_{snr} = -\frac{45}{2r_n^2}\frac{I}{a_n L_n F} \tag{5.189}$$

$$\frac{d\langle\bar{c}_{spr}\rangle}{dt} + \frac{30 D_{sn}}{r_k^2}\bar{c}_{spr} = \frac{45}{2r_p^2}\frac{I}{a_p L_p F} \tag{5.190}$$

$$\mathscr{D}\frac{dq_{lin}}{dt} = -B_2 q_{lin} - A_2 q_{lip} + A_3 B_2 \frac{I}{a_n L_n F} + A_2 B_3 \frac{I}{a_p L_p F}$$

$$\mathscr{D}\frac{dq_{lip}}{dt} = B_1 q_{lin} + A_1 q_{lip} - A_3 B_1 \frac{I}{a_n L_n F} - A_1 B_3 \frac{I}{a_p L_p F} \tag{5.191}$$

$$\frac{d\delta_f}{dt} = -\frac{j_s M_f}{\rho_f}. \tag{5.192}$$

3. The analytical expressions:

$$c_{sur,n} = \bar{c}_{sn} - \frac{r_n}{35 D_{sn}}\frac{I}{a_n L_n F} + \frac{8 r_n}{35}\bar{c}_{snr} \tag{5.193}$$

$$c_{sur,p} = \bar{c}_{sp} + \frac{r_p}{35 D_{sp}}\frac{I}{a_p L_p F} + \frac{8 r_p}{35}\bar{c}_{spr} \tag{5.194}$$

$$c_{lin} = c_{lip} + \frac{L_s\left(q_{lin} + q_{lip}\right)}{2 D_{ls}} \tag{5.195}$$

$$c_{lip} = c_{l0} + \alpha_{in} q_{lin} + \alpha_{ip} q_{lip} \tag{5.196}$$

$$c_{lmid} = c_{lin} - \frac{3 L_s}{8 D_{ls}} q_{lin} - \frac{L_s}{8 D_{ls}} q_{lip} \tag{5.197}$$

$$c_l\left(x = 0, t\right) = c_{lin}\left(t\right) + \frac{q_{lin}\left(t\right) L_n}{2 D_{ln}} \tag{5.198}$$

$$c_l\left(x = L, t\right) = c_{lip}\left(t\right) - \frac{q_{lip}\left(t\right) L_p}{2 D_{lp}} \tag{5.199}$$

$$j_s(x,t) = \frac{-j_{s0}}{F}\exp\left(-\frac{\alpha_c F}{R_g T}\left(\phi_{1n} - \phi_{2n} - U_s - j_n(x,t) R_f F\right)\right). \tag{5.200}$$

$$R_{sei} = R_{f_0} + \frac{\delta_f}{\kappa_{sei}} \tag{5.201}$$

4. Voltage drops due to the electrochemical processes:

$$V_{sc} = -\frac{IL_p}{2\sigma_{sp}} \tag{5.202}$$

$$\begin{aligned} V_{el} = &\ \phi_{lip}(t) + 2\theta ln\left(\tfrac{c_l(x=L,t)}{c_{lip}(x,t)}\right) - \tfrac{I(t)L_p}{2\kappa_{lp}} \\ &-\phi_{lin}(t) - 2\theta ln\left(\tfrac{c_l(x=0,t)}{c_{lin}(x,t)}\right) - \tfrac{I(t)L_n}{2\kappa_{ln}} \end{aligned} \tag{5.203}$$

$$V_0 = V_{0,p}\left(c_{sur,p}\right)|_{x=L} - V_{0,n}\left(c_{sur,n}\right)|_{x=0} \tag{5.204}$$

$$\begin{aligned} V_{rd} = &-\tfrac{2R_gT}{F}\sinh^{-1}\left(\tfrac{I}{2j_{p0}(c_{sur,p})a_pL_pF}\right)|_{x=L} \\ &-\tfrac{2R_gT}{F}\sinh^{-1}\left(\tfrac{I}{2j_{n0}(c_{sur,n})a_nL_nF}\right)|_{x=0} \end{aligned} \tag{5.205}$$

$$V_{sei} = \frac{I}{a_nL_n}R_{sei} \tag{5.206}$$

5. Finally, the cell voltage:

$$V_{\text{cell}} = V_0 + V_{sc} + V_{el} + V_{rd} + V_{sei} \tag{5.207}$$

Chapter 6
Key Applications of ROM

The reduced order model developed in the last chapter is solved for academic and realistic scenarios. For the latter, validation against constant-controlled constant current data as well as profiles resembling vehicle drive conditions are considered. For the former, special scenarios like cells with phase change electrodes and cycling studies are analyzed.[1]

6.1 Introduction

In this chapter, we will see the application of ROM to various scenarios—ranging from characterization of exotic features like multiphase coexistence to performance to drive cycle data. Due to the intricate coupling of the cell temperature through the OCV and the transport parameters, accurate estimation of temperature feeds into an accurate estimation of cell voltage, and hence the state and eventually health. Despite including intricate physical details, the electrochemical–thermal model with the PDE structure is not easily amenable for onboard implementation. Hence, a combination of model order reduction are attempted in various publications. Reference [7] uses of proper orthogonal decomposition method to obtain a reduced order finite volume representation of the rigorous electrochemical–thermal model. Reference [20] combine the lumped thermal model equation [4] with the single-particle model. Reference [37] also make use of the lumped thermal model by combining it with the average model which is a simplified form of the complete electrochemical model. Reference [6] make use of the single particle model with the lumped thermal balance equation to study the effect of the cooling fluid temperature and flow rate on the cell temperature. Reference [41] use a linear parameter varying model to reduce the order of the electrochemical model. Although good match with experimental data is seen,

[1]Figures and discussions reproduced with permissions from Elsevier.

K. S. Hariharan et al., *Mathematical Modeling of Lithium Batteries*,
Green Energy and Technology, https://doi.org/10.1007/978-3-319-03527-7_6

these and similar methods either use intricate numerical methods there by loosing out on physics, or retain elementary physics information.

The approach discussed in the last chapter differentiates by retaining the electrochemical basis. In this chapter, we will see this model applied for two different scenarios. The first discusses about the utility of ROM in representing realistic—experimental and drive cycle—data. As a demonstration, our research work on an NCA/C cell is discussed. In the next section, the focus is on the some atypical experimental response of cells with electrodes that exhibit multiphase coexistence. For this case, published experimental data of LFP/C cell is considered.

6.2 Representation of Experimental and Drive Cycle Data

The lithium-ion cell considered in the present section has lithium nickel cobalt aluminium oxide (LNCAO) composite as the positive electrode material and graphite (C) as the negative electrode. The electrochemical and thermal equations detailed in the previous section are solved in Simulink with a two-way coupling. The results obtained from the model are discussed in this section.

6.2.1 Details of Experiments

The experiments are conducted on a commercial lithium-ion cell with LNCAO as the positive electrode and a carbon-based negative electrode. To obtain charge–discharge data the cells are cycled using Solartron 1470E cell test system. The cycler is a multichannel (8 channels), 10 V (0.1% range), 4A (0.1% range) device (SH 642). It has 1470E (multi-channel potentiostats), 1455 (FRA), 14703A (temperature measurement option), and 14702A (auxiliary voltage channels). Three electrode experiments are performed on the commercially available LNCAO using lithium metal as a reference electrode. The charge discharge is carried out at a constant current rate of 0.01C. The cut off potential for charge and discharge is maintained at 4.2 V and 2.75 V, respectively. The solid phase diffusion coefficient of lithium in nickel cobalt aluminium oxide particle is estimated using electrochemical impedance spectroscopy (EIS) and galvanostatic intermittent titration technique (GITT). For the EIS experiment, the frequency range used is 1MHz–10mHz while the frequency resolution is 1 in 65,000,000. The current rate maintained during the GITT discharge is 0.01C.

6.2.2 Validation with CC Data

The charging is done using a constant current (CC)–constant voltage (CV) profile, while the discharge is constant current. During the operation, the cell is kept at isothermal conditions. Complete charge–discharge cycle is carried out under these conditions at various temperatures. At each of the temperature, CC–CV charge and CC discharge data is obtained at the current rates of 0.01, 0.1, 0.5, and 1C. The results from the validation at room temperature (298 K) and high temperature (333 K) are given in Figs. 6.1 and 6.2.

At 298 K (Fig. 6.1), a good match is observed at low rates except at the end of discharge for 0.01C, where the model under predicts the end capacity (~7%). The

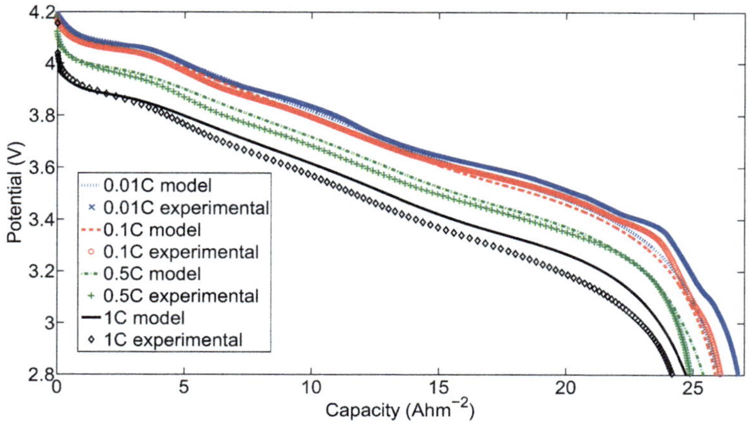

Fig. 6.1 Validation of model results with experimental data obtained at 298 K

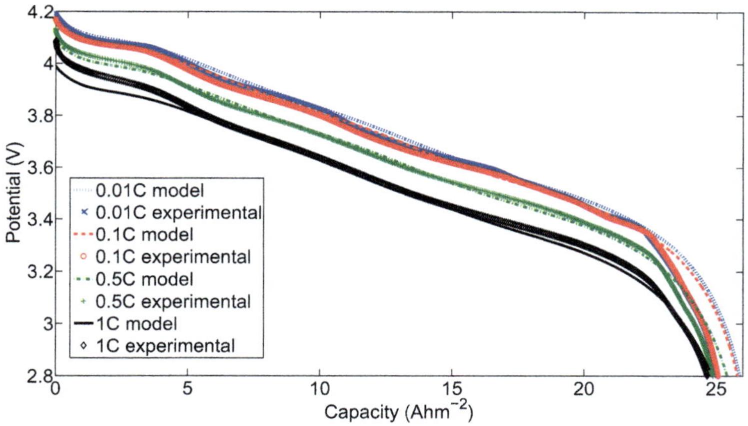

Fig. 6.2 Validation of model results with experimental data obtained at 333 K

match improves at 0.1C and 0.5C, while at 1C the model is seen to over predict the experiment result by $\sim$4%.

At high temperature (Fig. 6.2), the match between the model and experiments is substantially improved when compared to the room temperature data. Due to the increase in the transport properties at high temperature, the capacity drop between the 0.01C and 1C data is also observed to be lower.

6.2.3 Representation of Drive Cycle Data

Due to its simplified structure, the present ROM can be used to model transient and cyclic data in real time. In the present section, its capability is demonstrated by validating it with the following cyclic profiles. These profiles mimic the actual driving conditions of an EV.

6.2.3.1 Hybrid Pulse Power Characterization

The commercial LNCAO/C cell is subjected to an HPPC protocol. The temperature for this experiment is maintained at 333 K. The protocol includes a discharge pulse for 10 s at 5C then a 40 s rest (0C) followed by a charge pulse for 10 s at 3.75C. The cell is then discharged at 0.3 C for approximately 1225 s after it is rested for 3600 s to even out any gradients. This entire protocol is repeated until the cut-off voltage (2.75 V) is reached.

The ROM is used to model the behavior of LNCAO/C cell subjected to a HPPC cycle. The HPPC cycle includes a discharge (5C for 10 s)-rest (40 s)-charge (3.75C for 10 s). This pulse protocol is followed by discharge at ~0.3 C for ~1225 s and rest for 1 h after which the entire cycle is repeated. The experimental current data is used as the model input to obtain the output from ROM. The cell ambient is considered to be isothermal at 333 K as per the experimental procedure.

The discharge potential data from ROM is compared to the experimental cell response in Fig. 6.3. A good match is observed throughout the cycle (>96%) except at the end of discharge where the match is still >90%. The capability of ROM to capture the high rate, short time discharge and charge pulses effectively is demonstrated through the inset plots in the Fig. 6.3.

Due to its physics-based nature, ROM serves the dual capability of modeling real-time cell behavior and obtaining physical insights such as understanding the controlling processes involved in the cell operation. The HPPC protocol is used as a case study to demonstrate these capabilities of ROM as it involves all the conditions of operation a cell undergoes namely charge, discharge, rest, and high temperature operation. The contribution to the cell voltage from the various electrochemical processes, occurring within the cell, during a single discharge–rest–charge protocol

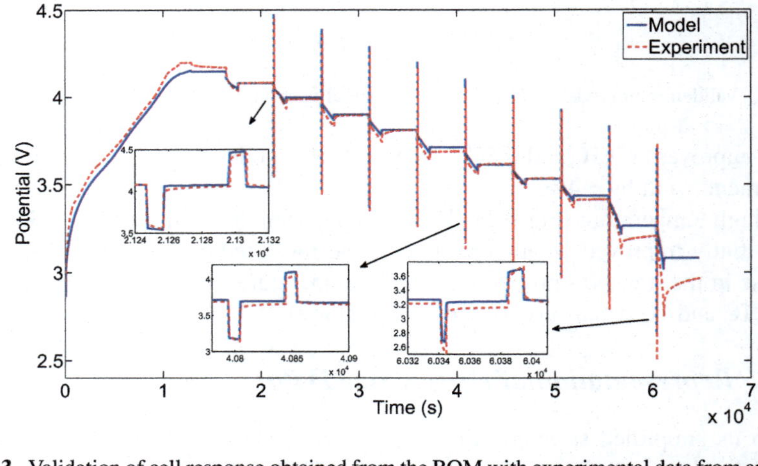

Fig. 6.3 Validation of cell response obtained from the ROM with experimental data from an HPPC cycle. Conditions used are $T_0 = 333$ K and $h = 1000$ Wm^{-2}K^{-1}

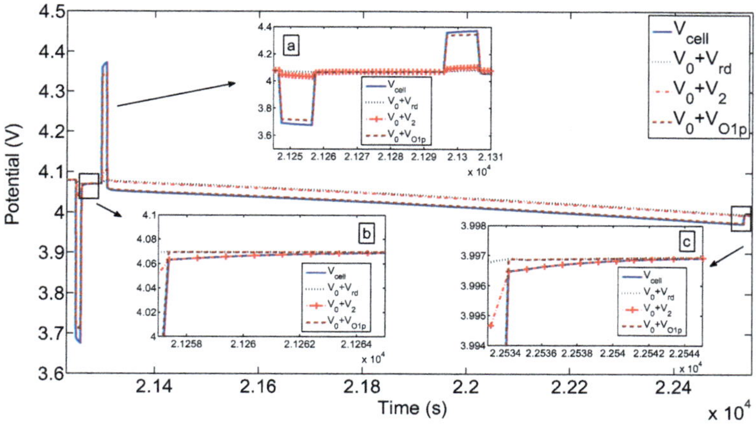

Fig. 6.4 Analysis of the contribution of reaction diffusion (V_{rd}), electrolyte (V_2) and Ohmic potentials (V_{o1p}) to the cell voltage during an HPPC operation. Conditions used are $R_f = 0$, $T_0 = 333$ K and $h = 1000$ Wm^{-2}K^{-1}

is analyzed. These physical insights obtained from ROM can be used to understand the internal dynamics governing the cell behavior and therefore provides a handle with which to tune the cell design to obtain desired response.

The details are plotted in Fig. 6.4. Each of the sources of potential drop within the LNCAO/C cell (that is, reaction diffusion (V_{rd}), electrolyte (V_2) and Ohmic potentials (V_{o1p})) are added individually to the OCV of the cell. The cell voltage (V_{cell}) is compared to these to identify the controlling/limiting process in the cell. To obtain the contributions of only the physical processes involved in cell operation the contact resistance is assumed to be negligible in this case. Owing to the low conductivity of the positive electrode, the Ohmic potential drop is a significant contributor (refer inset (a)) to cell voltage during high rate operation (both charge and discharge). The electrolyte phase potential is significant to a lower extent while the potential due to the electrochemical reaction is negligible during the charge and discharge pulses. As the cell enters the rest phase after the first discharge pulse (inset (b)), the electrolyte phase potential takes over as a controlling process. The term $V_0 + V_2$ is seen to match the V_{cell} at the beginning of the rest period. This is also observed at the end of the 0.3 C discharge (shown in inset (c)) when the cell again enters a rest phase.

6.2.3.2 Urban Dynamometer Driving Schedule

To obtain the cell performance for a realistic driving scenario the cell is subjected to a UDDS cycle. This cycle is suitable for the electric vehicles as it involves load profiles encountered in the urban areas. The definition of the UDDS is used to obtain a load profile of the discharge current for the cell. This includes transient pulses of varying

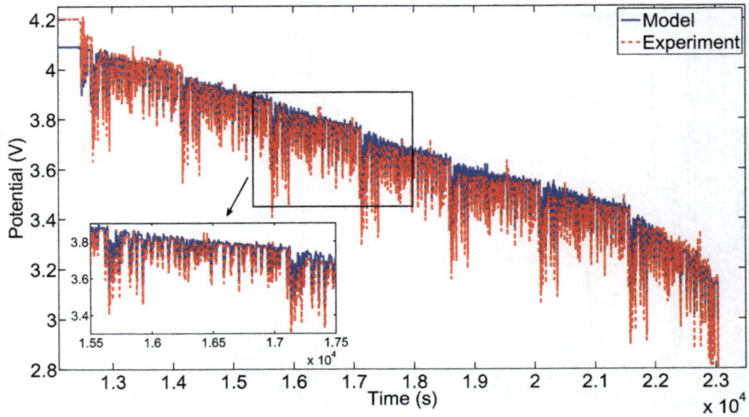

Fig. 6.5 Validation of cell response to UDDS cycle as obtained from the ROM with experimental data. Conditions used are $T_0 = 333$ K and $h = 1000\ \mathrm{Wm^{-2}K^{-1}}$

rates and the cycle is repeated after every 1484 s. The entire UDDS experiment is carried out isothermally at 298 K.

The applicability of ROM in predicting this highly transient data as experienced during real driving scenarios is demonstrated thus. The load profile in terms of current pulses is used as the input to the model. The obtained output of cell voltage is compared with the experimental response in Fig. 6.5. The model results match the experimental voltage values accurately with an error of 7% at points with maximum deviation.

6.2.4 *Time Taken*

One of the qualifying factors of a reduced order model is its computation speed. To demonstrate the capability of the present ROM, the time taken by it to simulate a single discharge of the LNCAO/C cell is compared to that of the full model in Table 6.1.

Table 6.1 A comparison of run times of different models

S.no	Model	Time taken
1	Real data	3492 s
2	P2D thermal model (in COMSOL)	22 s
3	Present ROM (thermal)	~2.63 s

Both the models are run on a system with a 3.4 GHz intel (R) i7 processor with a 3.48 GB memory. The cell behavior is modeled for a constant current discharge of 1C under isothermal conditions of 298 K. The experimental discharge takes 3492 s to reach the cut off voltage of 2.8 V. The P2D model solved using the finite element scheme in COMSOL takes around 22 s to model the complete discharge. ROM, however, completes the discharge in approximately 2.64 s. This estimate indicates that each data point of the discharge is available every $\sim 0.7 - 1$ millisecond. This exemplifies the significant advantage in speed provided by ROM for onboard application of physics-based models.

6.3 Parametric Sensitivity Analysis

After benchmarking the reduced order electrochemical thermal model with experimental data, it is used to study the LNCAO/C cell sensitivity to the following practical scenarios

1. Contribution to overall heat release
2. Effect of ambient temperature (T_0)
3. Effect of the rate of operation (I)
4. Effect of the cooling system (h).

6.3.1 Contribution to the Total Heat Release

To explore further the capability of ROM in obtaining the contribution of individual heat sources to total heat, the same are plotted as a function of the cell capacity in Fig. 6.6. For the present set of parameters for an LNCAO/C cell, the Ohmic heat contributed by the positive electrode is seen to be the highest. This is due to the low electronic conductivity of the LNCAO material. The irreversible heat due to the overpotential and the Ohmic heat from the negative electrode are less significant in comparison. The contribution of the reversible heat from both the electrodes is endothermic for the initial portion of the discharge cycle. The trends observed in the individual heat sources are reflected in the total heat from the cell. The total heat is also endothermic albeit to a lower extent owning to the balancing effect of exothermic Ohmic heat from the positive electrode. At around two-third discharge, the total heat becomes exothermic, gradually increases and peaks to around 2×10^4 Wm^{-3}.

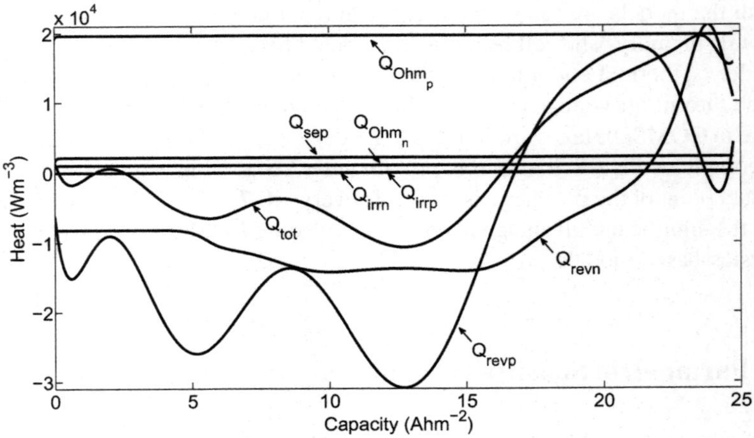

Fig. 6.6 Contribution of various heat sources to the total heat obtained from ROM

6.3.2 *Effect of Ambient Temperature*

To study the effect of ambient temperature on the cell response, the other two parameters namely, the rate of operation and the heat transfer coefficient are maintained at their nominal values, that is, 1C and $h = 0$.

The cell response is measured in terms of three variables: the cell potential (V), average change in temperature (T), and the total heat generated from the cell (Q_{tot}). In Figs. 6.7, 6.8 and 6.9, the effect of operating the LNCAO/C cell at low (273 K)

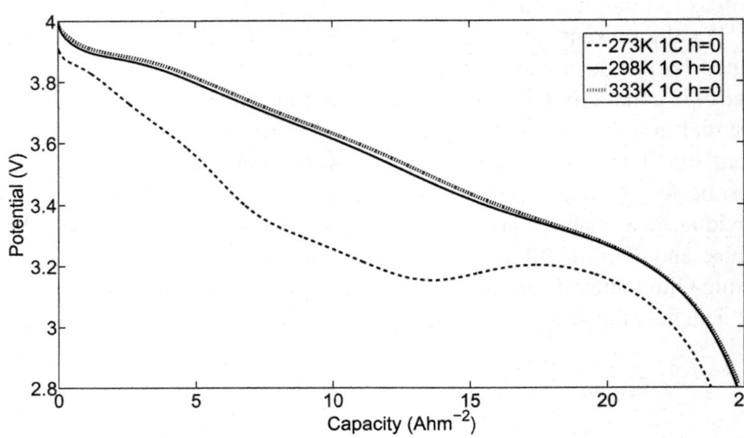

Fig. 6.7 Variation of cell potential at different temperatures

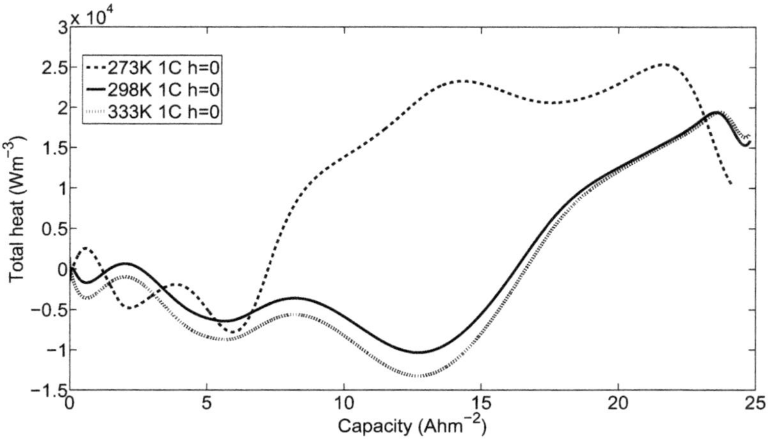

Fig. 6.8 Variation of total heat as a function of capacity at different temperatures

and high temperatures (333 K) is studied and compared to the operation at room temperature (298 K).

The cell is seen to discharge (Fig. 6.7) at the nominal rate to an end capacity of ∼25 Ahm^{-2}. The total heat (Fig. 6.8) is seen to be endothermic for almost two-third of the discharge cycle after which it becomes exothermic. A signature of this is observed in the cell average temperature (Fig. 6.9) wherein, the temperature steadily decreases until around a third of discharge capacity and then increases to approximately 1 K higher than the ambient.

The end capacity is observed to be high at higher temperatures and reduces with temperature (Fig. 6.7). The total heat on the other hand (Fig. 6.8) at 333 K is more endothermic than that at 298 K. After it becomes exothermic it crosses over the value at 298 K at the end of discharge. The cell temperature at 333 K mimics the behavior observed at room temperature (Fig. 6.9) except at the end of discharge where it ends closer to the ambient temperature. But as the ambient temperature is itself high, operation at 333 K has improved transport processes which lead to the higher capacity.

The discharge voltage at 273 K shows non-monotonic behavior (Fig. 6.7). The continually decreasing voltage sees a shift in the slope at ∼7 Ahm^{-2}. The decrease at the modified slope continues until ∼12 Ahm^{-2} where a dip in the voltage occurs. The discharge finally ends at capacity lower than that of higher temperatures. From the Fig. 6.8, the first anomaly in the discharge curve, the shift in the slope, occurs when the total heat from the system turns exothermic from an initial endothermic behavior. The heat then peaks at capacity corresponding to the dip in the discharge curve finally ending at a value lower than that seen for higher temperatures. The cell temperature, after a short period of initial decrease which corresponds to the system turning exothermic, steadily increases and reaches a higher value of ∼15 K.

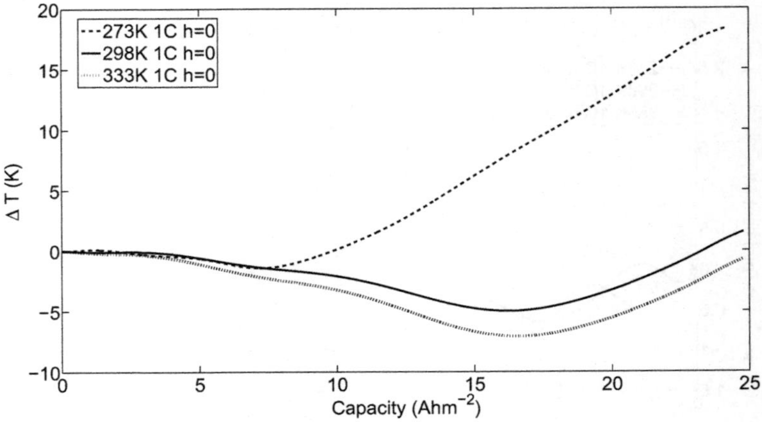

Fig. 6.9 Change in temperature as a function of capacity at different temperatures

6.3.3 *Effect of Rate*

The effect of discharge rate on the adiabatic operation of an LNCAO/C cell at room temperature is studied. The capacity drops substantially on increasing the discharge rate as observed in Fig. 6.10. The discharge potentials during isothermal operation are plotted as dashed lines for comparison. It can be observed that the capacity obtained during adiabatic operation of the cell is higher than that during isothermal operation. In addition, the extent of disparity in adiabatic and isothermal operation increases with increasing rate.

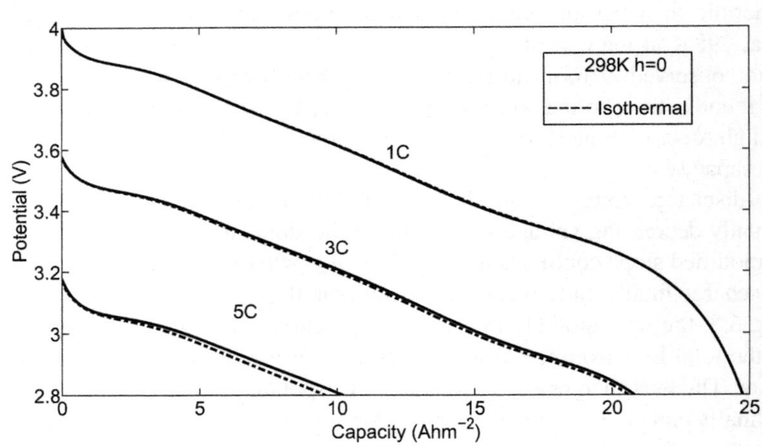

Fig. 6.10 Cell potential at different rates

The total heat corresponding to different discharge rates studied in Fig. 6.10 are plotted in Fig. 6.11. The trend at 1C as discussed earlier shifts from endothermic to exothermic heat. This trend is absent at higher rates of 3C and 5C. The total heat at high rates is always exothermic owing to an increase in the Ohmic heat which varies as the square of the applied current.

The result of the increased heat at higher rates is reflected in the temperature profiles during discharge. The temperature increases steadily at 3C with a sharper increase towards the end of discharge while at 5C the temperature increases rapidly by around 18 K (Fig. 6.12).

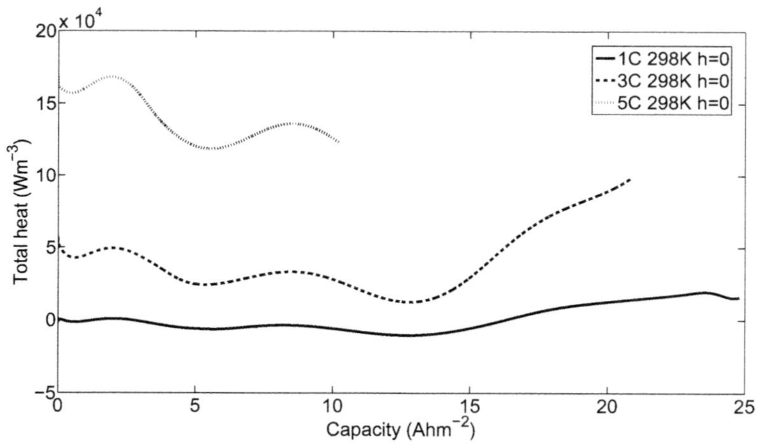

Fig. 6.11 Variation of total heat as a function of capacity at different rates

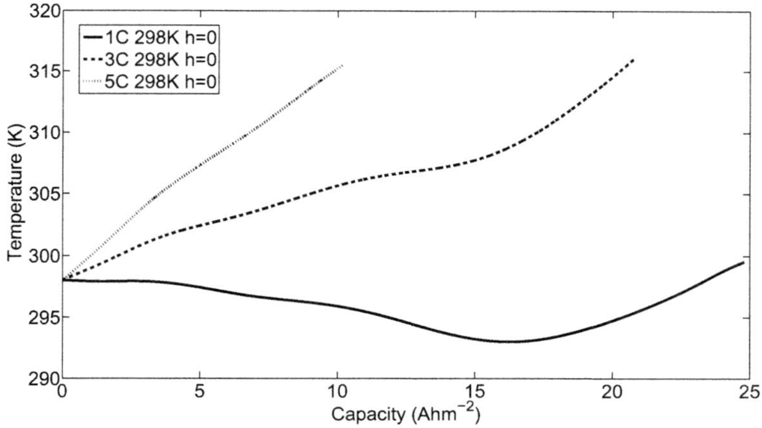

Fig. 6.12 Change in temperature as a function of capacity at different rates

6.3.4 Effect of Heat Transfer Coefficient

A study of varying the heat transfer coefficient is carried out in this section which can aid in designing a better thermal management for the LNCAO/C cells. In Fig. 6.13, the cell potential during discharge is studied at different values of the heat transfer coefficient ranging from adiabatic ($h = 0$ Wm^{2}K^{-1}) to isothermal ($h = 100$ Wm^{2}K^{-1}). Although the variation in the potential is negligible at room temperature and 1C, it is seen (from the inset in Fig. 6.13) that the discharge with $h = 1$ Wm^{2}K^{-1} shows higher end capacity. To understand the trend further, the temperature profiles are plotted in Fig. 6.14. For the adiabatic operation ($h = 0$ Wm^{2}K^{-1}), the temperature profile is as discussed in the previous figures wherein the temperature decreases initially and increases toward the end of discharge to reach a value slightly above the ambient. For the other extremum condition of heat transfer coefficient ($h = 100$ Wm^{2}K^{-1}) used, the temperature remains at its ambient value. For the intermediate value of $h = 1$ Wm^{2}K^{-1}, it is observed that the temperature initially decreases by less than 1K and then increases beyond its initial value. This increase in the cell temperature can lead to improved transport properties which are responsible for the increased end capacity at this value of h.

To explore the contribution of heat transfer coefficient in improving the discharge capacity further, the effect of h on the cell response at an ambient temperature of 273 K is studied in Fig. 6.15. The effect of lowering the heat transfer coefficient is clearly observed in the Fig. 6.15. In the beginning of discharge, the curve during adiabatic operation is seen to have lower voltage but towards the end of operation it provides the highest discharge capacity compared to other values of h. The reason for this trend can be inferred from the cell temperature data provided in the inset of the same figure. It can be seen that temperature for $h = 0$ increases rapidly to reach a value 15 K higher than the ambient. The temperature for operation at $h = 1$ also

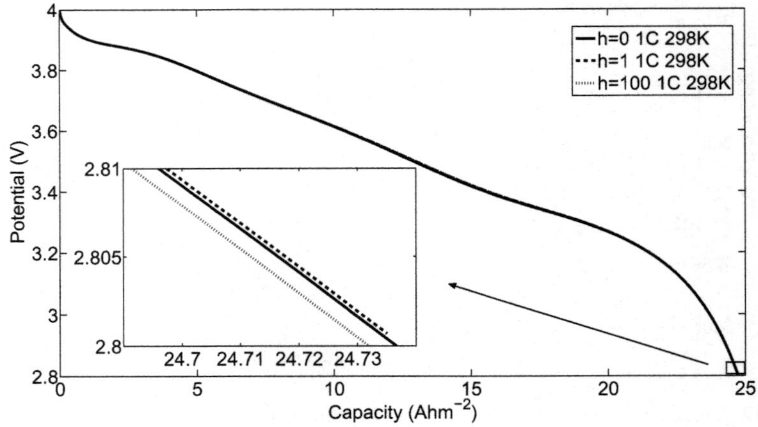

Fig. 6.13 Cell potential at different cooling conditions

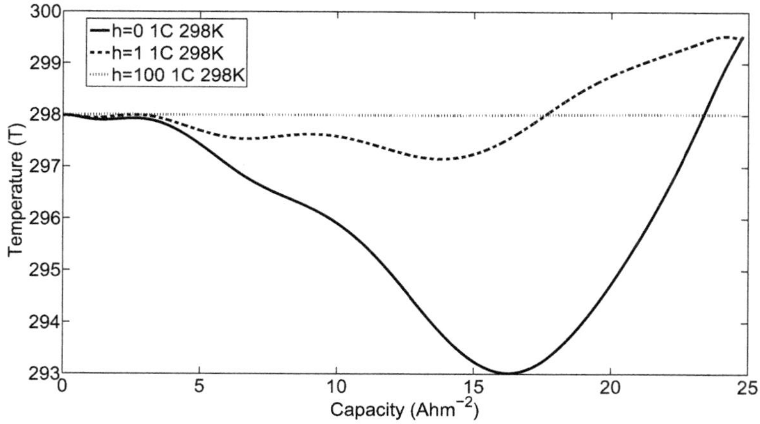

Fig. 6.14 Change in temperature as a function of capacity at different cooling conditions

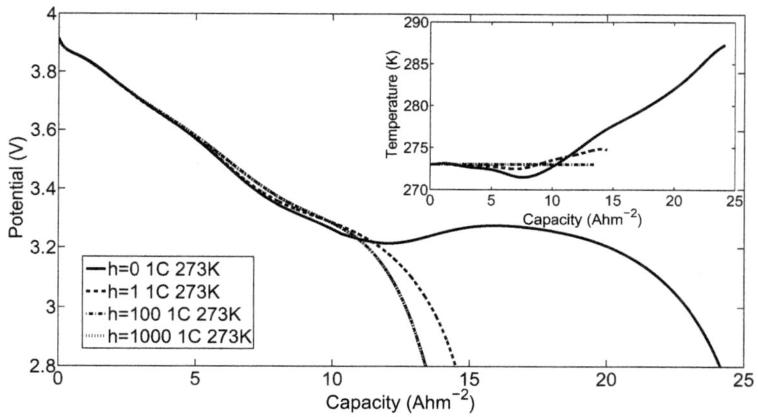

Fig. 6.15 Dependence of cell discharge on different cooling conditions at an ambient temperature of 273 K and 2C

increases but to a lower extent while for higher values of h the temperature remains at the ambient.

The advantage of the physics-based reduced order model for obtaining design insights is exemplified through this parametric sensitivity study. The model can be used to study varied scenarios which are experimentally difficult to realize and obtain physical insights on the lines of those obtained from the conventional electrochemical models. In addition, the reduced structure of the model enables onboard implementation and real-time data generation which are important features sought by device engineers.

6.4 Representation of Phase Change Electrodes

In the present section, results from the reduced order model (ROM) which includes the multiphase behavior of electrode materials and a kinetic SEI growth model to study the cycle life of cells, are presented. The first part deals with the discussion of fresh cell results followed by the cell aging studies.

6.4.1 Reduced Order Model for Phase Change System

To demonstrate its validity, the ROM for fresh cells (first cycle) is compared to experimental data from the literature. [45] report the galvanostatic charge–discharge curves at different rates for the LFP/C cell. For the present work, data points for two rates namely, C/10 and 1C are used for model validation. The comparison of model results to the experimental data is shown in Fig. 6.16. Good agreement between the experimental and simulated values is obtained, specially at the beginning and the end of the charge/discharge cycles. A relative mismatch is observed in the mid portion of the curves. This mismatch is seen to be higher during discharge than during the charge cycle. It is also found to be increasing with the rate but, the extent of the mismatch is found to be less than 1.8%. This substantiates the utility of this phase change-reduced order model at both low and nominal rates.

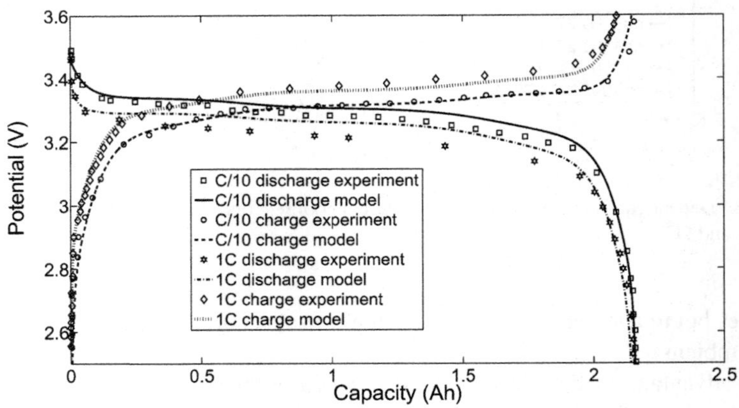

Fig. 6.16 The model results are validated with the experimental data reported in [45] at different rates for both charge and discharge cycles

6.4.2 Reduced Order Model for Phase Change System with Degradation

To study the degradation of the LFP/C cell, the side reaction rate is assumed to have finite value during the charging cycle. [43] report experimental cell response data for various cycles for an LFP/C cell. As the initial states ($SOC_0 = c_{10}/c_{1max}$) for these set of experiments are different, the values of $SOC_{n0} = 0.79$, $SOC_{p0} = 0.03$ are used for validating the cycling results. Results from the present model are compared with the first and the 310th experimental cycle data in the Fig. 6.17. An excellent match between the model and experimental results with $<1\%$ error is observed. This benchmarks the ROM approach used in the present work to model the degradation of a phase change cell and demonstrates the applicability of the model.

6.4.2.1 SEI Film Growth and Rate Dependence

As discussed in Sect. 5.3, the cell degrades due to side reactions occurring on the negative electrode surface during aging. The reaction consumes lithium ions causing capacity loss while the products form an SEI film on the active material particle surface increasing the cell resistance while in operation. The discharge curves at different cycles are plotted in Fig. 6.18. A loss in capacity is observed immediately after the first cycle indicating the formation of the SEI film. The film grows with time increasing the resistance which manifests as an initial voltage drop seen with increasing cycle number. The increase in resistance, however, lowers the overpotential for the side reaction which slows the growth of the film. This leads to a continuous albeit slower loss in capacity with increasing cycle numbers. The SEI film thickness

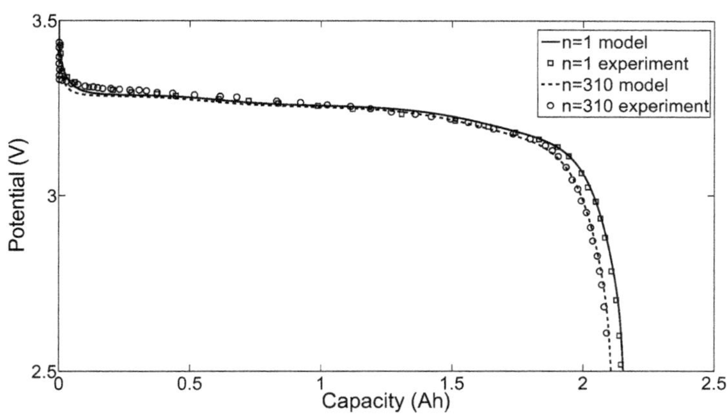

Fig. 6.17 The model results are validated with the cycle life experimental data reported in [43] at different discharge cycles

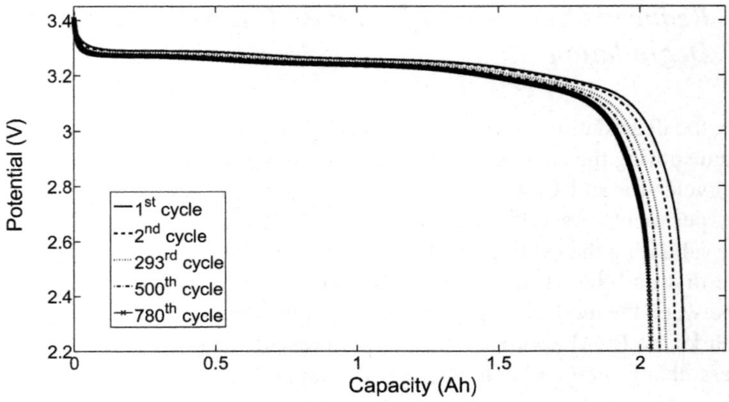

Fig. 6.18 The discharge curves of the LFP/C cell at different cycles

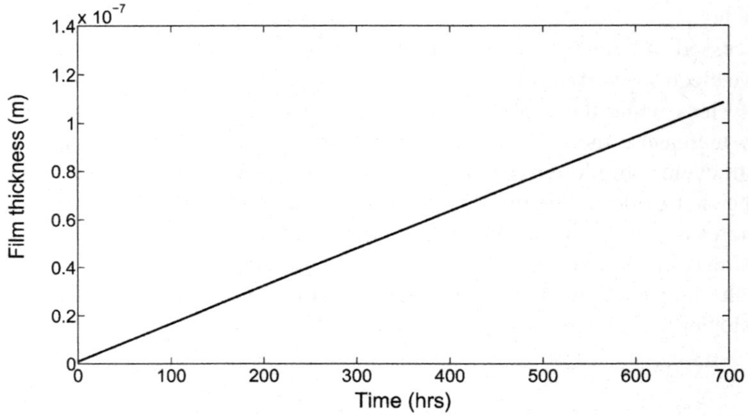

Fig. 6.19 The growth of the SEI film as a function of the cycling time in hours

as a function of time is presented in the Fig. 6.19. The film growth is found to be linearly increasing with the cycling time and hence, in extension, it is linear with the number of cycles, consistent with published results [5] in this domain. This provides a consistency check for the description of degradation used in the model.

The results from the ROM degradation model are used to study the rate dependence of the SEI film growth. For a fixed number of cycles, the increase in the SEI film thickness is plotted as a function of the current rate in Fig. 6.20. As the SEI formation occurs during the charge cycle, the charge rate is varied while keeping the discharge rate constant at 1C. From the Fig. 6.20, it can be observed that the SEI growth decreases with increase in the charge rate for the fixed number of cycles. At 5C, the SEI growth at the end of 5 cycles is almost equal to the growth observed in a single charge cycle at 1C.

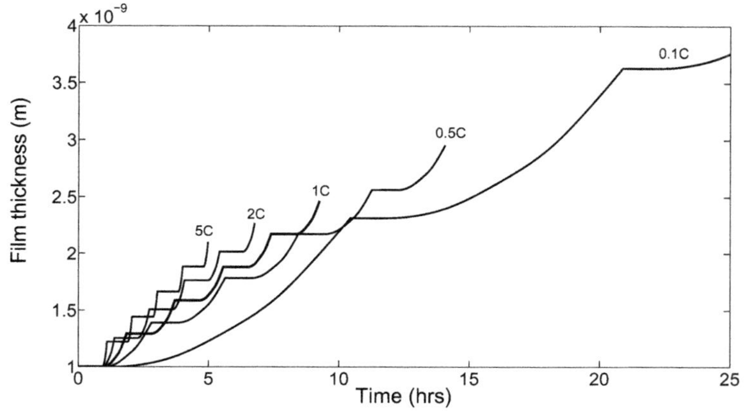

Fig. 6.20 The growth of the SEI film varying as a function of the current rate for a fixed number of cycles

This rate dependence of the SEI film thickness originates from an interplay of the side reaction rate and the time available for its growth. At higher current rates, a constant current charge cycle occurs in 1/5th of the time it would take at 1C. This gives lesser time for the side reaction to occur and hence leads to lower SEI film growth.

For high power applications low cell resistance is preferred. This translates to minimal SEI growth during the operation. While for high energy applications, to avoid continuous loss of lithium to the side reactions, formation of a stable SEI in the first cycle followed by a quick saturation in its growth is preferable. The insights from the present model's results can be used to design suitable charging profiles. For example, incorporating a higher rate constant current charging segment for a high power cell can lead to a lower SEI growth. While a lower rate constant current charging can lead to a rapidly saturating first cycle SEI growth for a high energy cell.

6.4.3 Charge–Discharge Asymmetry

Due to multiphase coexistence, an interesting feature of LFP cells is the charge–discharge asymmetry. In Fig. 6.16, an asymmetry in the end capacities (utilization) during charge and corresponding discharge cycle, is observed. To explore this observation further, rate capability results for the fresh LFP/C cell are studied in Fig. 6.21. Assuming a completely charged initial state, the cell discharge, at four different rates, are shown (solid lines) in Fig. 6.21. At the same rates, the cell response during charge, from a completely discharged state, are shown as dashed lines. At low rates, say C/10, the end capacity for discharge and charge are the same. While the end capacities do not vary significantly during discharge, the decrease in the end

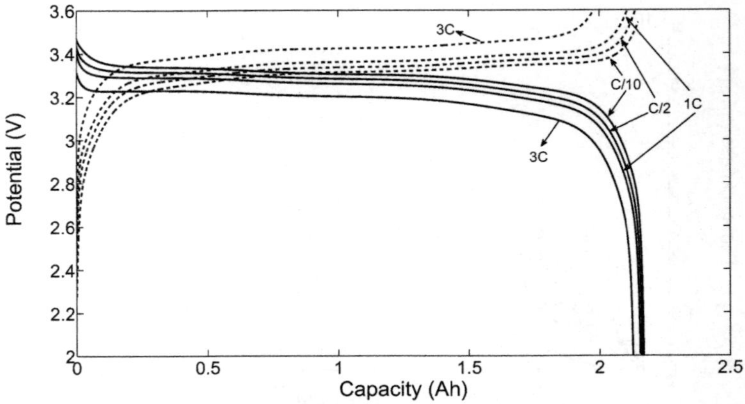

Fig. 6.21 Charge and discharge curves at different rates are plotted for the LFP/C cell under study. An asymmetry in the end capacities is observed

capacities during charge is quite significant. This gives rise to an asymmetry in the charge and discharge profiles. This asymmetric behavior is characteristic of the phase change electrode and has been reported previously [26, 44]. Thus, it can be seen that the proposed model can be used to characterize the observed atypical features of phase change cells.

It can be observed that the end capacity obtained via charging is lower than that obtained via discharging, at any particular rate. The difference in the end capacities is found to increase with increasing rate. The feature is observed in experiments also, and is attributed to the multiphase coexistence.

6.4.4 Path Dependence

The heterogeneous environment in the active material particle, due to the coexistence of two phases, affects the transport of lithium during insertion differently than during deinsertion. For example, for LFP during discharge, a lithium-rich phase (β) forms at the cost of the initial lithium-poor phase (α) and the reverse occurs during the charge cycle. As the diffusion coefficients are different for these two phases [50], lithium transport correspondingly takes shorter or longer depending on the phase it travels through.

A complete and accurate prediction of the path dependence of a cell is an important requirement for building the BMS [50]. This feature has earlier been explained using the shrinking core model [50] and the generalized moving boundary model [26]. These models have phase boundaries defined which specify the diffusion lengths involved in different paths. In the present work, in order to make the model amenable for multi-cycling studies, no explicit phase boundary is considered. It is, therefore, interesting to study if the path dependence can yet be captured by the present description of the phase change system.

In the fixed time protocol to study path dependence, the response of the cell subjected to two different paths is compared. In the first path (P1), the cell is discharged from a completely charged state for 2000 s. This forms the first stage of the path. In the second stage it is let to rest for 2 h. This long rest period ensures that the cell is equilibrated. In the third stage, post the resting, the cell is discharged until it reaches the cut-off voltage of 2 V. In the second path (P2), the difference is in the first stage, wherein, the cell is charged from a completely discharged state for 2000 s. The following two stages, the rest and discharge steps, remain same as those in P1. The details of the paths are summarized in Table 6.2. The cell potential obtained from the model, for the complete path, is plotted as a function of the cell capacity in Fig. 6.22. Irrespective of the initial stages, in both the cases, the cell is rested for two hours, yet, a clear difference is the end capacities is observed. The discharge curve of P2, with an initial charging stage, is found to have a higher end capacity than the other.

To understand the reason behind the path dependence, the average concentration of lithium in the active material particles, for both the paths, is plotted as a function of the cell capacity in Fig. 6.23. It can be seen from the figure, at the beginning of the stage 3, that the average concentration in the cell while undergoing P1 is higher than

Table 6.2 A summary of the various paths used in the present work, to study the path dependence in LFP/C cells

Fixed time protocol				
Path	Stage 1	Stage 2	Stage 3	Result
P1	1C, t = 2000 s	0C, t = 7200 s	1C, $V_{\text{cutoff}} = 2\,\text{V}$	Lower utilization
P2	−1C, t = 2000 s	0C, t = 7200 s	1C, $V_{\text{cutoff}} = 2\,\text{V}$	

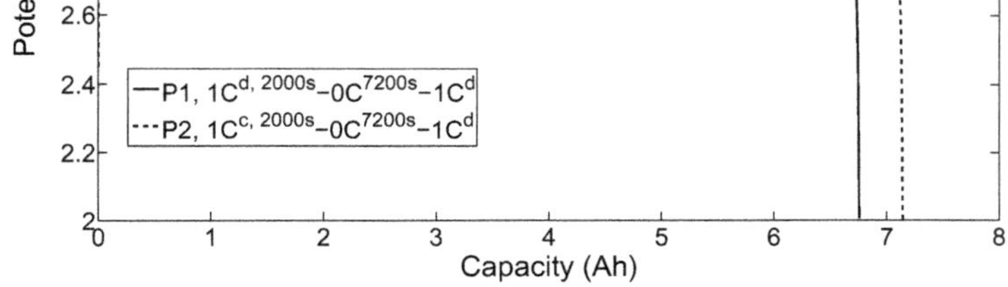

Fig. 6.22 The simulated response of the LFP/C cell to two different current paths

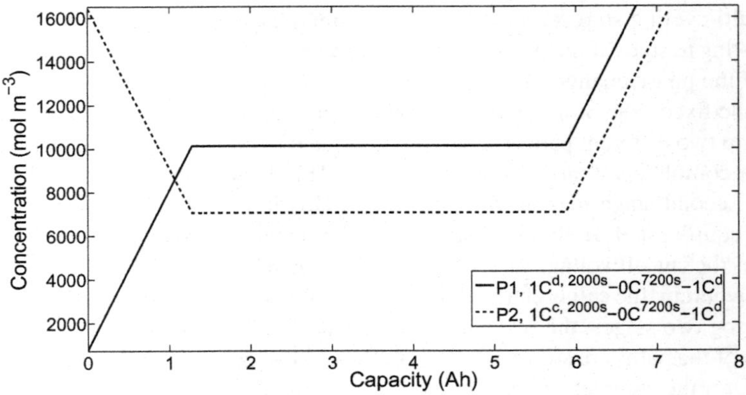

Fig. 6.23 The variation of the average concentration of the lithium in the positive electrode during the paths 1 and 2

when undergoing P2. This mismatch is the result of the asymmetry in the charge–discharge utilization during stage 1, as discussed in the Sect. 6.4.3. This, in turn, leads to an asymmetry in the end capacities originating from the initial paths taken.

Above discussion leads to following observations: Partial charging of phase change cells from a completely discharged state, before use (discharge), can lead to lower utilization. On the same lines paths 3 and 4 suggest that while charging, higher capacity can be attained if the preceding stage is charging. That is, lower utilization can occur when charging a partially discharged cell. The analysis of the model results show that the utilization can be maximized if similar stages (charge or discharge) succeed one another. The insights from this path dependence study can be used as operating guidelines for cells such as LFP/C that exhibit phase change.

6.5 Summary

This part deals with order reduction of the electrochemical–thermal (ECT) model. This important step makes the ECT amenable for onboard estimation. Due to this pivotal role there is recent research interest in reduced order models (ROM). The approach presented here uses volume averaging as a scaffolding to retain the physical significance of the equations. The equivalent physical variables that can be tracked by ODEs are identified and solved for. The ODEs are uncoupled and can be solved in a direct non-iterative fashion. On the base ROM framework additions involving energy balance, capacity loss and phase change electrodes are developed.

The ROM has the capability of obtaining the temperature profiles within the cell and the individual heat sources. The model is assessed in terms of its accuracy, comprehensiveness and speed in the last chapter. It is validated with experimental data at different rates at both high and low temperatures. A detailed study of cell behavior at different operating conditions in terms of temperature, rate, and cooling is carried out using the model. ROM accurately models the HPPC and UDDS cell discharges while providing understanding of the underlying phenomena. The multiphase coexistence within the active material, is introduced into ROM framework using a concentration-dependent diffusivity. The effect of SEI formation due to side reactions is implemented and validated with experimental data. It is demonstrated that the model can exhibit the observed experimental behavior of the phase change electrodes. It is also used to gain physical insight into various features of a phase change system such as charge–discharge asymmetry and path dependence. ROM provides data at less than a millisecond making it amenable for onboard use. The ROM discussed in these sections forms the template on which state estimators are integrated, as discussed subsequently.

Part III
State and Health Estimation

Chapter 7
Theoretical Framework for State Estimation

One of the most important functions of the battery management system is to accurately estimate the battery state using minimal onboard instrumentation. In this chapter, we present a recursive Bayesian filtering framework for onboard battery state estimation by assimilating measurables like cell voltage, current, and temperature with a physics-based model prediction. This framework can be numerically implemented using state-of-the-art filtering/data assimilation algorithms. We first develop a generic framework and then discuss some of the most popular algorithms for its implementation.

7.1 Introduction

In the previous chapters, our focus was on developing electrochemical thermal models of the Li-ion cell. Toward this effort, we first started with a discussion of the electrochemical processes in a Li-ion cell and subsequently presented a pseudo-two-dimensional electrochemical–thermal model of the Li-ion cell. However, the computational complexity of this model makes it difficult to implement on board the BMS. We have resolved this difficulty using volume averaging of the P2D-ECT model. The resultant reduced order model (ROM) retains the physical basis of the P2D-ECT model. In addition, the ROM is computationally simple enough to be implemented on board the BMS. However, the BMS with a standalone ROM running on it is not useful for our purpose because of following two sources of uncertainty.

- Execution of ROM requires specification of the initial SOC. However, the initial SOC of the battery is seldom known.
- Inherent model uncertainties due to poorly known parameters and unknown physics.

K. S. Hariharan et al., *Mathematical Modeling of Lithium Batteries*,
Green Energy and Technology, https://doi.org/10.1007/978-3-319-03527-7_7

These sources of uncertainty induce error in ROM predictions. This problem is further compounded as this error accumulates with time. As a result, the ROM prediction diverges from the true system response.

Fortunately, the BMS is equipped with onboard sensors that continuously monitor the battery parameters like cell voltage, current, and temperature. We can combine these sensor measurements with the ROM to improve the predictions. This process of combining sensor measurements with the mathematical model is known as data assimilation. Traditionally, the data assimilation is perceived as an error minimization process, where an optimization algorithm is used to minimize the (squared) error between model prediction and the measurements. However, this error minimization perspective fails to take into account internal physics of the system. In this chapter, we will introduce a recursive Bayesian framework that allows assimilation of the sensor measurements with the ROM while ensuring that the internal physics of the system is not violated. Note that this recursive Bayesian framework is known in the battery community as filtering and has been extensively used in the form of Kalman filter (and its variants) and particle filter. However these algorithms are used mainly from an error minimization perspective.

In this chapter, we show that these filtering methods are algorithms for implementing the recursive Bayesian framework at various levels of approximation. This also allows us to demonstrate physics-based coupling between a filtering algorithm and the mathematical model.

We will start our discussion with a detailed introduction to the Bayes theorem, which forms a basis for further development. Next, we will describe the recursive Bayesian framework for data assimilation. Subsequently, we will discuss various filtering algorithms for numerical implementation of the recursive Bayesian framework.

7.2 Bayes Theorem

Let us consider two random vectors $\boldsymbol{x}$ and $\boldsymbol{y}$ with respective probability density functions (probability density function (pdf)) $p(\boldsymbol{x})$ and $p(\boldsymbol{y})$. The joint pdf of these random vectors is given by $p(\boldsymbol{x}, \boldsymbol{y})$. For these random vectors, the conditional pdf is defined as

$$p(\boldsymbol{x}|\boldsymbol{y}) = \frac{p(\boldsymbol{x}, \boldsymbol{y})}{p(\boldsymbol{y})}, \tag{7.1}$$

where we assume that $p(\boldsymbol{y}) > 0$. Note that the conditional pdf $p(\boldsymbol{x}|\boldsymbol{y})$ is defined for a given value of $\boldsymbol{y}$, i.e., the distribution is a function of $\boldsymbol{y}$. We similarly define the conditional distribution $p(\boldsymbol{y}|\boldsymbol{x})$ as

$$p(\boldsymbol{y}|\boldsymbol{x}) = \frac{p(\boldsymbol{x}, \boldsymbol{y})}{p(\boldsymbol{x})}, \tag{7.2}$$

with $p(\boldsymbol{x}) > 0$. Using Eqs. 7.1 and 7.2, we obtain

$$p(\boldsymbol{x}, \boldsymbol{y}) = p(\boldsymbol{x}|\boldsymbol{y})\, p(\boldsymbol{y}) = p(\boldsymbol{y}|\boldsymbol{x})\, p(\boldsymbol{x}). \tag{7.3}$$

We can use symmetry of Eq. 7.3 to obtain

$$p(\boldsymbol{x}|\boldsymbol{y}) = \frac{p(\boldsymbol{y}|\boldsymbol{x})\, p(\boldsymbol{x})}{p(\boldsymbol{y})}. \tag{7.4}$$

The marginal pdf $p(\boldsymbol{y})$ is given by

$$\begin{aligned} p(\boldsymbol{y}) &= \int p(\boldsymbol{x}, \boldsymbol{y})\, dx \\ &= \int p(\boldsymbol{y}|\boldsymbol{x})\, p(\boldsymbol{x})\, dx. \end{aligned} \tag{7.5}$$

Using Eq. 7.5 in Eq. 7.4, we obtain a well-known statement of the Bayes theorem as

$$p(\boldsymbol{x}|\boldsymbol{y}) = \frac{p(\boldsymbol{y}|\boldsymbol{x})\, p(\boldsymbol{x})}{\int p(\boldsymbol{y}|\boldsymbol{x})\, p(\boldsymbol{x})\, dx}. \tag{7.6}$$

In Eq. 7.6, $p(\boldsymbol{x})$ is known as prior, $p(\boldsymbol{y}|\boldsymbol{x})$ is likelihood, $p(\boldsymbol{x}|\boldsymbol{y})$ is known as posterior and $\int p(\boldsymbol{y}|\boldsymbol{x})\, p(\boldsymbol{x})\, dx$ is known as the predictive probability distribution.

The Bayes theorem provides a rational and mathematically elegant framework for data assimilation. In this Bayesian framework, $\boldsymbol{x}$ is treated as a variable to be estimated, while $\boldsymbol{y}$ is treated as a data for estimating $\boldsymbol{x}$. As name suggests, the prior $p(\boldsymbol{x})$ probabilistically quantifies existing knowledge about $\boldsymbol{x}$. Generally, it is quite difficult to propose exact prior distribution and often approximations of "true" prior distribution are used in practice. To obtain these approximate prior distributions, various methods like maximum entropy priors, conjugate priors, and non-informative priors are proposed in the literature. Another method is to use a "hierarchical" approach, wherein a parametric prior distribution with unknown hyper-parameters is used. Subsequently these hyper-parameters are estimated using the Bayes theorem. In this book, we will not go into details of these methods. Readers are referred to Fink [12], Kass [40] and Paulo [38] for a more detailed discussion on the methods for specification of priors.

Likelihood $p(\boldsymbol{y}|\boldsymbol{x})$ is an important link between the Bayesian and traditional estimation methods like least-square and maximum likelihood estimation. In essence, the likelihood specifies probability of observing value $\boldsymbol{y}$ for the data given that $\boldsymbol{x}$ is a true estimate. Often, the likelihood is specified in terms of a likelihood function, which is defined as

$$\mathscr{L}(\boldsymbol{x}) = c\, p(\boldsymbol{y}|\boldsymbol{x}) \tag{7.7}$$

where the constant c can depend on $\boldsymbol{y}$ but not on $\boldsymbol{x}$. It may be noted that as a function of $\boldsymbol{y}$, $p(\cdot|\boldsymbol{x})$ is a probability measure, while as a function of $\boldsymbol{x}$, $p(\boldsymbol{y}|\cdot)$ is just a function which need not integrate to 1 and need not be additive. Importance of the likelihood function is characterize by the likelihood principle, which states that the inference from an experiment should be based only on the likelihood function of the observed data. The likelihood principle specifies the irrelevance of the events to the inference, except the observed one. One of the major advantages of the Bayesian inference is that it always satisfies the likelihood principle, whereas the other traditional inference techniques need not always satisfy the likelihood principle. It may be noted that the likelihood principle is only valid when the inference is about the same parameter and the parameter includes every unknown factor of the model. See Reid [31] for further details on the Likelihood.

The posterior distribution $p(\boldsymbol{x}|\boldsymbol{y})$ is at the core of the Bayesian framework. The posterior distribution characterizes the updated knowledge after addition of information available through the likelihood. The posterior distribution operates conditional on observations, thus automatically providing the inversion of probabilities while incorporating the likelihood principle. The posterior distribution also avoids the averaging over the unobserved events. In effect, prior distribution $p(\boldsymbol{x})$ quantifies the knowledge about $\boldsymbol{x}$ before observing the event $\boldsymbol{y}$, while the posterior distribution characterizes the knowledge after observing the event $\boldsymbol{y}$, hence the nomenclature. In case of the independent and identically distributed observations, the order in which observations are taken does not affect the posterior distribution. Also, the posterior distribution is unaffected if the inference is performed with one observation at a time or with a set of all the observations.

Predictive probability is a marginal probability of occurrence of an event $\boldsymbol{y}$. Since the predictive probability is just a normalizing constant, it does not play any significant role in the Bayesian inference. However, the predictive probability plays an important role in post-processing of the data, especially in evaluation of the posterior model. Let $\boldsymbol{y}$ denote an available observation which is used during the Bayesian inference while $\boldsymbol{y}^*$ denote a possible future observation. The probability of occurrence of $\boldsymbol{y}^*$ is given by

$$p\left(\boldsymbol{y}^*|\boldsymbol{y}\right) = \int p\left(\boldsymbol{y}^*|\boldsymbol{y}, \boldsymbol{x}\right) p(\boldsymbol{x}|\boldsymbol{y})\, d\boldsymbol{x}, \tag{7.8}$$

where $p(\boldsymbol{x}|\boldsymbol{y})$ is the posterior probability distribution. It may be noted that the domain of integral is all the possible values $\boldsymbol{x}$ can take after observing $\boldsymbol{y}$. In the rest of this chapter, we will extensively use the Bayes theorem and properties of its components to develop our recursive Bayesian framework and algorithms for its numerical implementation.

7.3 Recursive Bayesian Framework

The dynamical battery models developed in the previous chapters can be represented in a state-space form as

$$\mathbf{x}_{t+\Delta t} = f\left(\mathbf{x}_t, \mathbf{u}_t\right), \tag{7.9}$$

where f is a system model, $\mathbf{x} \in \mathscr{R}^d$ denotes model state, $\mathbf{u} \in \mathscr{R}^m$ are control inputs, t denotes a time instance and Δt is a model time step. To simplify the notation, we use time step indexing $\mathbf{x}_k$ in this chapter, where $\mathbf{x}_k = \mathbf{x}_t$ and $t = k\Delta t$. The model is integrated forward in time using a numerical integration scheme like Runge–Kutta and its variants.

The states $\mathbf{x}_k$ cannot be observed experimentally, and thus, are known as hidden states. However, we can experimentally observe the system- level properties that are a function of the hidden states. These properties are known as the observable or visible states. Experimental observation of the visible states are related to the hidden state as

$$\mathbf{y}_k = g(\mathbf{x}_k) + \boldsymbol{\varepsilon}, \tag{7.10}$$

where $g(\cdot)$ is an observation model and $\mathbf{y}_k$ denote measurements. $\boldsymbol{\varepsilon}$ is a zero-mean Gaussian noise that quantifies measurement uncertainty due to the finite sensor precision.

The state-space model in Eq. 7.9 is initialized from an initial condition $\mathbf{x_0}$. However, this initial condition is often unknown/poorly known, inducing uncertainty in $\mathbf{x_0}$. Uncertainty in the initial condition is specified using a probability distribution function $p\left(\mathbf{x_0}\right)$. The initial state uncertainty propagates forward in time, inducing uncertainty in the future states. This uncertainty propagation is modeled using a discrete-time stochastic process. This process is given by a conditional probability distribution

$$\boldsymbol{x}_k \sim p\left(\boldsymbol{x}_k | \boldsymbol{x}_{0:k-1}\right). \tag{7.11}$$

Equation (7.11) is a probabilistic counterpart of the state-space model (7.9). If the system is initialized from $\mathbf{x_0}$, and subsequently follows a path $\mathbf{x_1}, \mathbf{x_2}, \ldots, \mathbf{x_{k-1}}$, the Eq. 7.9 specifies probability of the system taking a value $\mathbf{x_k}$ at time instance k. Similarly, probabilistic counterpart of the measurement model is given by

$$\boldsymbol{y}_k \sim p\left(\boldsymbol{y}_k | \boldsymbol{x}_{0:k-1}, \boldsymbol{y}_{1:k-1}\right). \tag{7.12}$$

At an arbitrary time instance $T > k$, the Bayesian framework concerns an accurate estimation of $p\left(\mathbf{x}_{0:T} | \mathbf{y}_{1:k}\right)$. Using definition of the conditional probability, we obtain

$$p\left(\mathbf{x}_{0:T} | \mathbf{y}_{1:k}\right) = p\left(\mathbf{x}_{0:k} | \mathbf{y}_{1:k}\right) p\left(\mathbf{x}_{k+1:T} | \mathbf{x}_{0:k}, \mathbf{y}_{1:k}\right). \tag{7.13}$$

Thus, the Bayesian framework is implemented in two parts. In the first part, measurements are assimilated with the model predictions to obtain $p\left(\mathbf{x}_{0:k} | \mathbf{y}_{1:k}\right)$. The

Bayesian framework uses Bayes theorem to assimilate the monitored parameters with the predicted state. At an arbitrary time step k, the Bayes theorem is given by

$$p\left(\mathbf{x}_{0:k}|\mathbf{y}_{1:k}\right) \propto p\left(\mathbf{y}_{1:k}|\mathbf{x}_{0:k}\right) p\left(\mathbf{x}_{0:k}\right), \tag{7.14}$$

where $p\left(\mathbf{x}_{0:k}\right)$ is known as prior, $p\left(\mathbf{y}_{1:k}|\mathbf{x}_{0:k}\right)$ is the likelihood, and $p\left(\mathbf{x}_{0:k}|\mathbf{y}_{1:k}\right)$ is known as the posterior probability.

Though explicit, analytical solution of Eq. 7.14 is intractable except for a linear model with Gaussian uncertainties. For nonlinear models with non-Gaussian uncertainties, solution of Eq. 7.14 requires implementation of a numerical sampling technique. However, these sampling techniques are computationally expensive, rendering implementation of Eq. 7.14 intractable online.

To ensure online tractability, we make two assumptions in the Eq. 7.14. First, we approximate the stochastic process in the Eq. 7.11 as a Markov process. A Markov process specifies that the state at time k is dependent only on the state at time $k-1$, and independent on the history. Probabilistically, the Markov property is specified as

$$p\left(\boldsymbol{x}_k|\boldsymbol{x}_{0:k-1}\right) = p\left(\boldsymbol{x}_k|\boldsymbol{x}_{k-1}\right). \tag{7.15}$$

Using the definition of the conditional probability, the prior in Eq. 7.14 is given by

$$p\left(\mathbf{x}_{0:k}\right) = p\left(\boldsymbol{x}_0\right) \prod_{i=1}^{k} p\left(\boldsymbol{x}_i|\boldsymbol{x}_{0:i-1}\right). \tag{7.16}$$

Using the Markov approximation, the prior simplifies to

$$p\left(\mathbf{x}_{0:k}\right) = p\left(\boldsymbol{x}_0\right) \prod_{i=1}^{k} p\left(\boldsymbol{x}_i|\boldsymbol{x}_{i-1}\right). \tag{7.17}$$

As we cannot directly observe the states $\boldsymbol{x}$, stochastic model defined by Eq. 7.17 is known as hidden Markov model.

Second, the measurement uncertainty is assumed independent and identically distributed. Using this approximation, the likelihood in Eq. 7.14 is simplified as

$$p\left(\mathbf{y}_{1:k}|\mathbf{x}_{0:k}\right) = \prod_{i=1}^{k} p\left(\mathbf{y}_i|\mathbf{x}_i\right). \tag{7.18}$$

Using Eqs. (7.17)–(7.18) in the Bayes theorem Eq. (7.14), we obtain

$$p\left(\mathbf{x}_{0:k}|\mathbf{y}_{1:k}\right) \propto p\left(\mathbf{x}_0\right) \prod_{i=1}^{k} p\left(\mathbf{y}_i|\mathbf{x}_i\right) p\left(\mathbf{x}_i|\mathbf{x}_{i-1}\right). \tag{7.19}$$

Formulation Eq. 7.19 is known as Recursive Bayesian Framework (recursive Bayesian framework (RBF)).

For the Lithium-ion cell state estimation, we are interested in estimating the SOC at the current time instance by assimilating the data till the current time instance. This problem is popularly known as the Bayesian filtering In the probabilistic framework, the Bayesian filtering problem can be considered as an estimation of the conditional probability $p\left(\boldsymbol{x}_k | \boldsymbol{y}_{1:k}\right)$. As we demonstrate in the following, Bayesian filtering is a special case of the RBF formulation.

Using the definition of the marginal probability distribution, we obtain

$$p\left(\boldsymbol{x}_k | \boldsymbol{y}_{1:k}\right) = \int p\left(\boldsymbol{x}_{0:k} | \boldsymbol{y}_{1:k}\right) d\boldsymbol{x}_{0:k-1}. \tag{7.20}$$

Using the RBF formulation Eq. 7.19 in Eq. 7.20, we obtain

$$p\left(\boldsymbol{x}_k | \boldsymbol{y}_{1:k}\right) \propto \int p\left(\mathbf{x}_0\right) \prod_{i=1}^{k} p\left(\mathbf{y}_i | \mathbf{x}_i\right) p\left(\mathbf{x}_i | \mathbf{x}_{i-1}\right) d\boldsymbol{x}_{0:k-1} \tag{7.21}$$

$$= p\left(\mathbf{y}_k | \mathbf{x}_k\right) \int p\left(\mathbf{x}_0\right) p\left(\mathbf{x}_1 | \mathbf{x}_0\right) \prod_{i=2}^{k} p\left(\mathbf{x}_i | \mathbf{x}_{i-1}\right) p\left(\mathbf{y}_{i-1} | \mathbf{x}_{i-1}\right) d\boldsymbol{x}_{0:k-1}. \tag{7.22}$$

Using the Markov property, integral in Eq. 7.22 is given by

$$p\left(\mathbf{x}_k | \mathbf{y}_{1:k-1}\right) = \int p\left(\mathbf{x}_0\right) p\left(\mathbf{x}_1 | \mathbf{x}_0\right) \prod_{i=2}^{k} p\left(\mathbf{x}_i | \mathbf{x}_{i-1}\right) p\left(\mathbf{y}_{i-1} | \mathbf{x}_{i-1}\right) d\boldsymbol{x}_{0:k-1}. \tag{7.23}$$

Using Eq. 7.23 in Eq. 7.22, we obtain

$$p\left(\mathbf{x}_k | \mathbf{y}_{1:k}\right) \propto p\left(\mathbf{y}_k | \mathbf{x}_k\right) p\left(\mathbf{x}_k | \mathbf{y}_{1:k-1}\right), \tag{7.24}$$

which allows recursion.

The solution of Bayesian filtering Eq. 7.24 is implemented in the following two steps:

- Prediction Step: The predictive probability distribution $p\left(\mathbf{x}_k | \mathbf{y}_{1:k-1}\right)$ is estimated by evaluating the integral in Eq. 7.23.
- Update Step: At time instance k, the measurement $\boldsymbol{y}_k$ is assimilated to update the state using Eq. 7.24.

Unfortunately, except for the simplest linear models, both these steps are analytically intractable. State-of-the-art filtering algorithms, starting from the Kalman filter to the more advanced particle filter, numerically implement these two steps at various level of approximations.

7.3.1 *Kalman Filter*

Kalman filter [24] is a simplest and most widely used algorithm for numerical implementation of the Bayesian filter. The Kalman filter is a closed-form solution of the Bayesian filter for a linear model with Gaussian uncertainty. A linear state-space and measurement model is given by

$$\begin{aligned} \boldsymbol{x}_k &= \boldsymbol{A}_{k-1}\boldsymbol{x}_{k-1} + \boldsymbol{q}_{k-1} \\ \boldsymbol{y}_k &= \boldsymbol{H}_k\boldsymbol{x}_k + \boldsymbol{\varepsilon} \end{aligned} \tag{7.25}$$

where $\boldsymbol{A}_{k-1}$ is a transition matrix, $\boldsymbol{q}_{k-1}$ represents model structural uncertainty, $\boldsymbol{H}_k$ is the measurement model matrix and $\boldsymbol{\varepsilon}$ represents the measurement uncertainty.

In the Kalman filter, $\boldsymbol{q}_{k-1}$ and $\boldsymbol{\varepsilon}$ are assumed zero-mean Gaussian random variables with known variance. In particular, the following choices are used

$$\begin{aligned} \boldsymbol{q}_{k-1} &\sim \mathcal{N}\left(\boldsymbol{0}, \boldsymbol{Q}_{k-1}\right) \\ \boldsymbol{\varepsilon} &\sim \mathcal{N}\left(\boldsymbol{0}, \boldsymbol{R}\right). \end{aligned} \tag{7.26}$$

Model, Eq. 7.25, is initialized from an uncertain initial condition $\boldsymbol{x}_0$ with the probability distribution

$$p\left(\boldsymbol{x}_0\right) = \mathcal{N}\left(\boldsymbol{\mu}_0, \boldsymbol{P}_0\right). \tag{7.27}$$

Using these choices, Markov property and the linearity of the model (Eq. 7.25), the probabilistic state-space form is given by

$$\begin{aligned} p\left(\boldsymbol{x}_k|\boldsymbol{x}_{k-1}\right) &= \mathcal{N}\left(\boldsymbol{x}_k; \boldsymbol{A}_{k-1}\boldsymbol{x}_{k-1}, \boldsymbol{Q}_{k-1}\right) \\ p\left(\boldsymbol{y}_k|\boldsymbol{x}_k\right) &= \mathcal{N}\left(\boldsymbol{y}_k; \boldsymbol{H}_k\boldsymbol{x}_k, \boldsymbol{R}_k\right). \end{aligned} \tag{7.28}$$

Using this probabilistic model, we derive the prediction and update steps of the Bayesian filter.

7.3.1.1 Prediction Step

The predictive probability distribution in Eq. 7.22 is given by

$$\begin{aligned} p\left(\boldsymbol{x}_k|\boldsymbol{y}_{1:k-1}\right) &= \int p\left(\boldsymbol{x}_k, \boldsymbol{x}_{k-1}|\boldsymbol{y}_{1:k-1}\right) d\boldsymbol{x}_{k-1} \\ &= \int p\left(\boldsymbol{x}_k|\boldsymbol{x}_{k-1}\right) p\left(\boldsymbol{x}_{k-1}|\boldsymbol{y}_{1:k-1}\right) d\boldsymbol{x}_{k-1}. \end{aligned} \tag{7.29}$$

Using the probabilistic state-space form (Eq. 7.28), we obtain

$$p\left(\boldsymbol{x}_k|\boldsymbol{y}_{1:k-1}\right) = \int \mathcal{N}\left(\boldsymbol{x}_k;\, \boldsymbol{A}_{k-1}\boldsymbol{x}_{k-1},\, \boldsymbol{Q}_{k-1}\right) \mathcal{N}\left(\boldsymbol{x}_{k-1};\, \boldsymbol{\mu}_{k-1},\, \boldsymbol{P}_{k-1}\right) d\boldsymbol{x}_{k-1} \tag{7.30}$$

Using properties of the multivariate Gaussian distribution, analytical solution of the integral in Eq. 7.30 is available. Using the analytical solution, the predictive probability distribution is given by

$$p\left(\boldsymbol{x}_k|\boldsymbol{y}_{1:k-1}\right) = \mathcal{N}\left(\boldsymbol{x}_k;\, \boldsymbol{\mu}_k,\, \boldsymbol{P}_k\right), \tag{7.31}$$

where

$$\begin{aligned} \boldsymbol{\mu}_k &= \boldsymbol{A}_{k-1}\boldsymbol{\mu}_{k-1} \\ \boldsymbol{P}_k &= \boldsymbol{A}_{k-1}\boldsymbol{P}_{k-1}\boldsymbol{A}_{k-1}^T. \end{aligned} \tag{7.32}$$

7.3.1.2 Update Step

Use Eqs. 7.28 and 7.32 in the Bayesian filter formulation Eq. 7.24 to obtain

$$p\left(\mathbf{x}_k|\mathbf{y}_{1:k}\right) \propto \mathcal{N}\left(\boldsymbol{y}_k;\, \boldsymbol{H}_k\boldsymbol{x}_k,\, \boldsymbol{R}_k\right) \mathcal{N}\left(\boldsymbol{x}_k;\, \boldsymbol{\mu}_k,\, \boldsymbol{P}_k\right). \tag{7.33}$$

When both the prior and likelihood are Gaussian, posterior distribution obtained using the Bayes theorem is also Gaussian. Thus, the posterior distribution in Eq. 7.33 is given by

$$p\left(\mathbf{x}_k|\mathbf{y}_{1:k}\right) = \mathcal{N}\left(\boldsymbol{x}_k;\, \boldsymbol{\mu}_k^a,\, \boldsymbol{P}_k^a\right), \tag{7.34}$$

where

$$\boldsymbol{\mu}_k^a = \boldsymbol{\mu}_k + \mathcal{K}\left(\boldsymbol{y}_k - \boldsymbol{H}_k\boldsymbol{\mu}_k\right) \tag{7.35}$$

$$\boldsymbol{P}_k^a = \left(\boldsymbol{I} - \mathcal{K}\boldsymbol{H}_k\right)\boldsymbol{P}_k. \tag{7.36}$$

$\mathcal{K}$ is known as the Kalman gain, which is given as

$$\mathcal{K} = \boldsymbol{P}_k\boldsymbol{H}_k^T\left(\boldsymbol{H}_k\boldsymbol{P}_k\boldsymbol{H}_k^T + \boldsymbol{R}\right)^{-1}. \tag{7.37}$$

The Kalman filter implementation is summarized in the Algorithm 1.

The Kalman filter provides closed-form solution for the linear dynamical and measurement models with Gaussian uncertainties. However, real-life systems are seldom linear. Next, we review extended Kalman filter and unscented Kalman filter algorithms that extends the Kalman filter algorithm for nonlinear systems.

Algorithm 1 Kalman Filter

1: Initialize with $\mathbf{x}_0$, $\mathbf{P}_0$, $\mathbf{Q}_0$. Specify the measurement uncertainty $\mathbf{R}$.
2: Prediction step at time instance k:
  1: Obtain the mean using $\boldsymbol{\mu}_k = \boldsymbol{A}_{k-1}\boldsymbol{\mu}_{k-1}$
  2: Obtain the covariance using $\boldsymbol{P}_k = \boldsymbol{A}_{k-1}\boldsymbol{P}_{k-1}\boldsymbol{A}_{k-1}^T$
3: Update step at time instance k:
  1: Calculate the Kalman gain using $\mathscr{K} = \boldsymbol{P}_k\boldsymbol{H}_k^T\left(\boldsymbol{H}_k\boldsymbol{P}_k\boldsymbol{H}_k^T + \boldsymbol{R}\right)^{-1}$
  2: Obtain the updated mean using $\boldsymbol{\mu}_k^a = \boldsymbol{\mu}_k + \mathscr{K}\left(\boldsymbol{y}_k - \boldsymbol{H}_k\boldsymbol{\mu}_k\right)$
  3: Obtain the updated covariance using $\boldsymbol{P}_k^a = \left(\boldsymbol{I} - \mathscr{K}\boldsymbol{H}_k\right)\boldsymbol{P}_k$
4: Set $\boldsymbol{\mu}_k = \boldsymbol{\mu}_k^a$, $\boldsymbol{P}_k = \boldsymbol{P}_k^a$

7.3.2 Extended Kalman Filter

The extended Kalman filter uses first-order Taylor series approximation to linearize the dynamical and measurement models. Subsequent, numerical implementation is similar to the Kalman filter algorithm.

Consider a nonlinear model

$$\boldsymbol{y} = f\left(\boldsymbol{x}\right), \tag{7.38}$$

with uncertainty in $\boldsymbol{x}$ given by $\boldsymbol{x} \sim N\left(\boldsymbol{\mu}, \boldsymbol{P}\right)$. We can use a re-parametrization $\boldsymbol{x} = \boldsymbol{\mu} + \delta\boldsymbol{x}$ such that $\delta\boldsymbol{x} \sim N\left(\boldsymbol{0}, \boldsymbol{P}\right)$. Use this re-parametrization in a Taylor series approximation to obtain

$$f\left(\boldsymbol{x} + \delta\boldsymbol{x}\right) \approx f\left(\boldsymbol{\mu}\right) + \boldsymbol{A}_x\delta\boldsymbol{x} + \text{higher order terms} \tag{7.39}$$

where $\boldsymbol{A}_x$ is a Jacobian matrix given by

$$\boldsymbol{A}_{x\,i,j} = \frac{\partial_i f\left(\boldsymbol{x}\right)}{\partial x_j}. \tag{7.40}$$

Neglecting higher order terms in the Taylor series approximation (Eq. 7.39), probability distribution of $\boldsymbol{y}$ can be approximated as a Gaussian

$$\boldsymbol{y} \sim N\left(\boldsymbol{\mu}_y, \boldsymbol{P}_y\right). \tag{7.41}$$

Mean of the distribution is given by

$$\boldsymbol{\mu}_y = E\left(f\left(\boldsymbol{\mu}\right) + \boldsymbol{A}_x\delta\boldsymbol{x}\right) \tag{7.42}$$

$$= f\left(\boldsymbol{\mu}\right). \tag{7.43}$$

Similarly, the covariance matrix is given by

$$\boldsymbol{P}_y = E\left((f(\boldsymbol{\mu}) + \boldsymbol{A}_x\delta\boldsymbol{x} - f(\boldsymbol{\mu}))(f(\boldsymbol{\mu}) + \boldsymbol{A}_x\delta\boldsymbol{x} - f(\boldsymbol{\mu}))^T\right) \tag{7.44}$$
$$= \boldsymbol{A}_x E(\delta\boldsymbol{x}(\delta\boldsymbol{x})^T)\boldsymbol{A}_x^T \tag{7.45}$$
$$= \boldsymbol{A}_x \boldsymbol{P} \boldsymbol{A}_x^T. \tag{7.46}$$

These Gaussian approximations are used to derive the extended Kalman filter.

Consider a state-space model

$$\boldsymbol{x}_k = f(\boldsymbol{x}_{k-1}) + \boldsymbol{q}_{k-1} \tag{7.47}$$
$$\boldsymbol{y}_k = g(\boldsymbol{x}_k) + \boldsymbol{\varepsilon}, \tag{7.48}$$

where $\boldsymbol{q}_{k-1} \sim N(\boldsymbol{0}, \boldsymbol{Q}_{k-1})$, $\boldsymbol{\varepsilon} \sim N(\boldsymbol{0}, \boldsymbol{R})$, $f(\cdot)$ is a nonlinear system model and $g(\cdot)$ is a nonlinear measurement model. Let $\boldsymbol{A}_{xk-1}$ and $\boldsymbol{H}_{xk}$ are Jacobian matrices of $f(\cdot)$ and $g(\cdot)$ respectively. Using the first-order Taylor series expansion (Eq. 7.39) and the associated Gaussian approximation (Eq. 7.41), we obtain the following probabilistic state-space model

$$p(\boldsymbol{x}_k|\boldsymbol{x}_{k-1}) = \mathcal{N}\left(\boldsymbol{x}_k; f\left(\boldsymbol{\mu}_{k-1}\right), \boldsymbol{A}_{xk-1}\boldsymbol{P}_{k-1}\boldsymbol{A}_{xk-1}^T + \boldsymbol{Q}_{k-1}\right) \tag{7.49}$$
$$p\left(\boldsymbol{y}_k|\boldsymbol{x}_k\right) = \mathcal{N}\left(\boldsymbol{y}_k; g\left(\boldsymbol{\mu}_k\right), \boldsymbol{H}_{xk}\boldsymbol{P}_k\boldsymbol{H}_{xk}^T + \boldsymbol{R}\right) \tag{7.50}$$

Probability distributions defined in Eq. 7.50 are used to derive the prediction and update steps of the extended Kalman filter.

7.3.2.1 Prediction Step

Using Eq. 7.50, the predictive probability distribution is approximated as

$$p\left(\boldsymbol{x}_k|\boldsymbol{y}_{1:k-1}\right) = \mathcal{N}\left(\boldsymbol{x}_k; \boldsymbol{\mu}_k, \boldsymbol{P}_k\right) \tag{7.51}$$

where

$$\boldsymbol{\mu}_k = f\left(\boldsymbol{\mu}_{k-1}\right) \tag{7.52}$$
$$\boldsymbol{P}_k = \boldsymbol{A}_{xk-1}\boldsymbol{P}_{k-1}\boldsymbol{A}_{xk-1}^T + \boldsymbol{Q}_{k-1}. \tag{7.53}$$

7.3.2.2 Update Step

Using the Gaussian approximation of Eq. 7.50 in Eq. 7.24, we obtain

$$p\left(\boldsymbol{x}_k|\boldsymbol{y}_{1:k}\right) = \mathcal{N}\left(\boldsymbol{x}_k; \boldsymbol{\mu}_k^a, \boldsymbol{P}_k^a\right), \tag{7.54}$$

where

$$\boldsymbol{\mu}_k^a = f\left(\boldsymbol{\mu}_{k-1}\right) + \mathscr{K}\left(\boldsymbol{y}_k - g\left(\boldsymbol{\mu}_k\right)\right) \tag{7.55}$$
$$\boldsymbol{P}_k^a = \left(\boldsymbol{I} - \mathscr{K}\boldsymbol{H}_{xk}\right)\boldsymbol{P}_k. \tag{7.56}$$

Similar to the Kalman filter, $\mathscr{K}$ is a Kalman gain given by

$$\mathscr{K} = \boldsymbol{P}_k \boldsymbol{H}_{xk}^T \left(\boldsymbol{H}_{xk}\boldsymbol{P}_k\boldsymbol{H}_{xk}^T + \boldsymbol{R}\right)^{-1}. \tag{7.57}$$

The extended Kalman filter algorithm is implemented using Algorithm 2.

Algorithm 2 Extended Kalman Filter

1: Initialize with $\mathbf{x}_0$, $\mathbf{P}_0$, $\mathbf{Q}_0$. Specify the measurement uncertainty $\mathbf{R}$.
2: Prediction step at time instance k:
  1: Obtain the Jacobians $\boldsymbol{A}_{xk-1}$ and $\boldsymbol{H}_{xk-1}$
  2: Obtain the mean using $\boldsymbol{\mu}_k = f\left(\boldsymbol{\mu}_{k-1}\right)$
  3: Obtain the covariance using $\boldsymbol{P}_k = \boldsymbol{A}_{xk-1}\boldsymbol{P}_{k-1}\boldsymbol{A}_{xk-1}^T + \boldsymbol{Q}_{k-1}$
4: Update step at time instance k:
  1: Calculate the Kalman gain using $\mathscr{K} = \boldsymbol{P}_k\boldsymbol{H}_{xk}^T\left(\boldsymbol{H}_{xk}\boldsymbol{P}_k\boldsymbol{H}_{xk}^T + R\right)^{-1}$
  2: Obtain the updated mean using $\boldsymbol{\mu}_k^a = f\left(\boldsymbol{\mu}_{k-1}\right) + \mathscr{K}\left(\boldsymbol{y}_k - g\left(\boldsymbol{\mu}_k\right)\right)$
  3: Obtain the updated covariance using $\boldsymbol{P}_k^a = \left(\boldsymbol{I} - \mathscr{K}\boldsymbol{H}_{xk}\right)\boldsymbol{P}_k$
4: Set $\boldsymbol{\mu}_k = \boldsymbol{\mu}_k^a$, $\boldsymbol{P}_k = \boldsymbol{P}_k^a$

7.3.3 Unscented Kalman Filter

Unscented Kalman Filter (unscented Kalman filter (UKF)) uses the unscented transformation [23] of the system for solution of RBF framework. The unscented transformation is a method of approximating statistics of a Gaussian random variable that undergoes a nonlinear transformation. To calculate the statistics, the method uses a set of deterministic realizations of the random variable, known as the sigma vectors. For an n-dimensional random variable $\mathbf{x}$ with mean $\bar{\mathbf{x}}$ and covariance $\mathbf{P}_\mathbf{x}$, $2n+1$ sigma-vectors are given by

$$\mathbf{X} = \left[\bar{\mathbf{x}} \quad \bar{\mathbf{x}} \pm \sqrt{(n+\lambda)\mathbf{P}_\mathbf{x}}\right], \tag{7.58}$$

where λ is a scaling parameter. The sigma-vectors are associated with the weights given by $W_0 = \lambda/(n+\lambda)$ and $W_i = 1/\{2(n+\lambda)\}$. The nonlinear transformation is applied to each sigma-vector, and the mean and covariance of the transformed variable are estimated using weighted mean and covariance of the transformed sigma-vectors. The resultant estimation of the mean and covariance are third-order accurate for the Gaussian inputs and at least second-order accurate for the non-Gaussian inputs.

The idea of unscented transform is used by UKF for solution of the RBF framework. The UKF algorithm is initialized using sigma-vectors of the initial state. For the forecast step, every sigma-vector is propagated forward in time by solving the system model Eq. 7.9. Mean and covariance of the predicted state are calculated at each time step. For the update step, well-known analysis equations of the KF are applied to update the predicted mean and the covariance. To estimate $p\left(\mathbf{x}_{k+1:T} | \mathbf{x}_{0:k}, \mathbf{y}_{1:k}\right)$, the prediction step is repeated for each time step. Algorithm 3 summarizes the UKF implementation.

Algorithm 3 Unscented Kalman Filter

1: Initialize with $\mathbf{x}_0, \mathbf{P}_0$. Calculate sigma-vectors $\mathbf{X}_0$ and weights $\mathscr{W}^m$, $\mathscr{W}^c$.
2: Prediction step:
  1: Calculate $\mathbf{X}_k = f(\mathbf{X}_{k-1})$
  2: Calculate observation sigma-vector $\mathbf{Y}_k = g(\mathbf{X}_k)$
  3: Calculate relevant statistics using

$$\overline{\mathbf{x}_k^-} = \sum_i \mathscr{W}_i^m \mathbf{X}_{i,k}$$
$$\overline{\mathbf{P}_k^-} = \sum_i \mathscr{W}_i^c \left[\mathbf{X}_{i,k} - \overline{\mathbf{x}_k^-}\right]\left[\mathbf{X}_{i,k} - \overline{\mathbf{x}_k^-}\right]^T$$
$$\overline{\mathbf{y}_k^-} = \sum_i \mathscr{W}_i^m \mathbf{Y}_{i,k}$$

1: Data assimilation:
  1: Calculate covariance matrices

$$\mathbf{P}_{\mathbf{y}_k\mathbf{y}_k} = \sum_i \mathscr{W}_i^c \left[\mathbf{Y}_{i,k} - \overline{\mathbf{y}_k^-}\right]\left[\mathbf{Y}_{i,k} - \overline{\mathbf{y}_k^-}\right]^T$$
$$\mathbf{P}_{\mathbf{x}_k\mathbf{y}_k} = \sum_i \mathscr{W}_i^c \left[\mathbf{X}_{i,k} - \overline{\mathbf{x}_k^-}\right]\left[\mathbf{Y}_{i,k} - \overline{\mathbf{y}_k^-}\right]^T$$

  2: Calculate Kalman gain as $\mathscr{K} = \mathbf{P}_{\mathbf{x}_k\mathbf{y}_k}\mathbf{P}_{\mathbf{y}_k\mathbf{y}_k}^{-1}$
  3: Update mean and covariance as

$$\mathbf{x}_k = \mathbf{x}_k^- + \mathscr{K}\left(\mathbf{y}_k - \overline{\mathbf{y}_k^-}\right)$$
$$\mathbf{P}_k = \mathbf{P}_k^- - \mathscr{K}\mathbf{P}_{\mathbf{y}_k\mathbf{y}_k}\mathscr{K}^T$$

7.3.4 Particle Filter

Particle Filter (particle filter (PF)) is a data assimilation algorithm for a nonlinear state-space model with non-Gaussian uncertainties. PF uses (Monte Carlo (MC)) sampling for a solution of the RBF formulation, and as such, is popularly known as the Sequential Monte Carlo (SMC) method [1]. Any SMC method is a variant of an importance sampling algorithm to efficiently sample from an intractable PDF. For sampling from an arbitrary PDF $p(\mathbf{x})$, the SMC method first collects an ensemble of particles from an appropriately chosen importance sampling distribution $\{\mathbf{x}^{(i)}\} \sim q(\mathbf{x})$. The PDF $p(\mathbf{x})$ is then approximated as

$$p(\mathbf{x}) \approx \sum_{i=1}^{P} \omega^{(i)} \delta\left(\mathbf{x}^{(i)} - \mathbf{x}\right), \tag{7.59}$$

where $\delta(\cdot)$ is a delta function and $\omega^{(i)}$ are the importance weights given by

$$\omega^{(i)} = \frac{p\left(\mathbf{x}^{(i)}\right)}{q\left(\mathbf{x}^{(i)}\right)}. \tag{7.60}$$

The particle filter is derived by using the importance sampling recursively. The recursion is obtained by applying importance sampling to the posterior distribution $p\left(\boldsymbol{x}_{0:k} | \boldsymbol{y}_{1:k}\right)$ in Eq. 7.14. Using the importance sampling, the posterior distribution is approximated as

$$p\left(\boldsymbol{x}_{0:k} | \boldsymbol{y}_{1:k}\right) \approxeq \sum_{i=1}^{P} \omega_k^i \delta\left(\boldsymbol{x}_k^i - \boldsymbol{x}_k\right), \tag{7.61}$$

where the weights are given by

$$\omega_k^i = \frac{p\left(\boldsymbol{x}_{0:k}^i | \boldsymbol{y}_{1:k}\right)}{q\left(\boldsymbol{x}_{0:k}^i | \boldsymbol{y}_{1:k}\right)}. \tag{7.62}$$

Here, $q\left(\boldsymbol{x}_{0:k} | \boldsymbol{y}_{1:k}\right)$ is an importance distribution. Using Markov property with the Bayes theorem in Eq. 7.62, we obtain

$$p\left(\boldsymbol{x}_k^i | \boldsymbol{y}_{1:k}\right) \propto p\left(\boldsymbol{y}_k | \boldsymbol{x}_k^i\right) p\left(\boldsymbol{x}_k^i | \boldsymbol{x}_{k-1}^i\right) p\left(\boldsymbol{x}_{0:k-1}^i | \boldsymbol{y}_{1:k-1}\right) \tag{7.63}$$

Similarly using the Bayes theorem, importance distribution is given by

$$q\left(\boldsymbol{x}_{0:k}^i | \boldsymbol{y}_{1:k}\right) \propto q\left(\boldsymbol{x}_k^i | \boldsymbol{x}_{1:k-1}^i, \boldsymbol{y}_{1:k}\right) q\left(\boldsymbol{x}_{1:k-1}^i | \boldsymbol{y}_{1:k-1}^i\right). \tag{7.64}$$

Using Eqs. 7.63–7.64 in Eq. 7.62, we obtain

$$\begin{aligned} \omega_k^i &\propto \frac{p\left(\boldsymbol{y}_k | \boldsymbol{x}_k^i\right) p\left(\boldsymbol{x}_k^i | \boldsymbol{x}_{k-1}^i\right) p\left(\boldsymbol{x}_{0:k-1}^i | \boldsymbol{y}_{1:k-1}\right)}{q\left(\boldsymbol{x}_k^i | \boldsymbol{x}_{1:k-1}^i, \boldsymbol{y}_{1:k}\right) q\left(\boldsymbol{x}_{1:k-1}^i | \boldsymbol{y}_{1:k-1}^i\right)} \\ &\propto \omega_{k-1}^i \frac{p\left(\boldsymbol{y}_k | \boldsymbol{x}_k^i\right) p\left(\boldsymbol{x}_k^i | \boldsymbol{x}_{k-1}^i\right)}{q\left(\boldsymbol{x}_k^i | \boldsymbol{x}_{1:k-1}^i, \boldsymbol{y}_{1:k}\right)}, \end{aligned} \tag{7.65}$$

where

$$\omega_{k-1}^{i} \propto \frac{p\left(\boldsymbol{x}_{0:k-1}^{i}|\boldsymbol{y}_{1:k-1}\right)}{q\left(\boldsymbol{x}_{1:k-1}^{i}|\boldsymbol{y}_{1:k-1}^{i}\right)}. \tag{7.66}$$

The sampling importance resampling (Sampling importance resampling (SIR)) algorithm [18] is a most widely used PF method. The algorithm uses the prior PDF as an importance sampling distribution, thus

$$q\left(\boldsymbol{x}_{1:k-1}^{i}|\boldsymbol{y}_{1:k-1}^{i}\right) = p\left(\boldsymbol{x}_{k}^{i}|\boldsymbol{x}_{k-1}^{i}\right). \tag{7.67}$$

The SIR algorithm is used to obtain prediction and update steps of the particle filter.

7.3.4.1 Prediction Step

The predictive probability distribution is given by

Algorithm 4 Particle Filter

1: Create initial ensemble $\{\mathbf{x}_0^i \sim p(\mathbf{x})\}$. Initialize weights to $\omega_0^i = 1/P$.
2: Prediction step: Calculate $\{\mathbf{x}_k^i = f(\mathbf{x}_{k-1}^i, \mathbf{u}_{k-1})\}$. Retain the weights unchanged.
3: Data assimilation: Calculate the updated weights using $w_k^i = \omega_{k-1}^{(i)} p\left(\mathbf{y}_k|\mathbf{x}_k^{(i)}\right)$. For a Gaussian likelihood, the updated weights are given by

$$w_k^i = \exp\left(-\frac{\left(\mathbf{y}_k - \mathbf{x}_k^i\right)^2}{2\sigma_y^2}\right) \tag{7.72}$$

4: Normalize the weights using

$$\omega_k^i = \frac{w_k^i}{\sum_{j=1}^{P} w_k^j}. \tag{7.73}$$

5: Calculate the effective number of samples using

$$P^{eff} = \frac{1}{\sum_{j=1}^{P}\left(\omega_k^i\right)^2}. \tag{7.74}$$

6: If $P^{eff} < P_T$, where P_T is threshold number of samples, resample using
   1: Construct CDF $c_i = c_{i-1} + \omega_k^i$ with $c_1 = 0$.
   2: Draw indices $i_r \sim \{c_i\}$
   3: Assign sample $\mathbf{x}_k^i = \mathbf{x}_k^{i_r}$
   4: Assign weight $\omega_k^i = 1/P$.

$$p\left(\boldsymbol{x}_k|\boldsymbol{y}_{1:k-1}\right) = p\left(\boldsymbol{x}_k|\boldsymbol{x}_{k-1}\right) p\left(\boldsymbol{y}_{1:k-1}|\boldsymbol{x}_{1:k-1}\right) \tag{7.68}$$

$$= \sum_{i=1}^{P} \omega_{k-1}^i \delta\left(\boldsymbol{x}_k^i - \boldsymbol{x}_k\right) \tag{7.69}$$

$$= \sum_{i=1}^{P} \omega_{k-1}^i \delta\left(f\left(\boldsymbol{x}_{k-1}^i\right) - \boldsymbol{x}_k\right) \tag{7.70}$$

Thus, the prediction step is implemented by simulating the model for each particle $\boldsymbol{x}_{k-1}^i$, while the weights remain unchanged.

7.3.4.2 Update Step

Using the choice of the importance sampling distribution Eq. 7.67 in Eq. 7.65, the weight update is given by

$$\omega_k^{(i)} = \omega_{k-1}^{(i)} p\left(\mathbf{y}_k|\mathbf{x}_k^{(i)}\right). \tag{7.71}$$

Thus, the update step of the particle filter is implemented by updating the weights using Eq. 7.71.

However after a few update steps, the PF suffers from the sample degeneracy problem, where most of the importance weights are close to zero. To avoid sample degeneracy, the particles are resampled when the effective number of samples crosses a predefined threshold. During the resampling step, the current ensemble of particles is replaced by a new set which is obtained by drawing particles from the current ensemble with probability proportional to the respective weight. The weights of the resampled particles are reassigned to $1/P$. To establish the predictive capability, the prediction step is repeated at each time step without the weight update. The PF is implemented using Algorithm 4.

Chapter 8
Theoretical Framework for Health Estimation Using Machine Learning

Real-time prediction of Remaining Useful Life (RUL) is an essential feature of a robust battery management system (BMS). However, due to the complex nature of the battery degradation, physics-based degradation modeling is often infeasible. Data-driven approaches provide an alternative when physics-based modeling is infeasible. In this chapter, we investigate some of the most popular machine learning-based data-driven approaches used by the Lithium-ion battery community. The chapter first introduces basic concepts of classification and regression, followed by a generic framework for its solution. Finally, we introduce some machine learning algorithms for the solution of this generic framework.

8.1 Introduction

In this book, we started our journey of Lithium-ion battery modeling from fully physics-based electrochemical thermal models. Subsequently, we presented a model order reduction methodology to develop an efficient and accurate ROM for the Lithium-ion battery. We subsequently explored the role of data in Lithium-ion battery modeling and developed a methodology to combine sensor measurements with the physics-based models. Our journey, which started from purely physics-based models and moved toward fusion models, will end in this chapter with a purely data-driven approach. In particular, we will explore some of the purely data-driven machine learning approaches that are popular in the Lithium-ion battery community. However, before proceeding further, we must consider scenarios where purely data-driven approaches can provide visible benefits over the physics-based approaches.

The data-driven approaches are especially beneficial in the following scenarios:

- When some physics of the system is poorly known or completely unknown.

K. S. Hariharan et al., *Mathematical Modeling of Lithium Batteries*,
Green Energy and Technology, https://doi.org/10.1007/978-3-319-03527-7_8

- When the physics-based model is computationally very expensive and cannot be used for online real-time applications.

Both these conditions are satisfied for state-of-health (SOH) estimation of the Lithium-ion batteries. Health monitoring based on machine learning tools is gaining importance for SOH estimation in recent years [57]. Prominent among these are artificial neural networks (ANN) [2, 25, 57] (reviewed in [42]), support vector machines (SVM), and Gaussian process estimators. These machine learning approaches are primarily used for classification and regression. Classification is a useful tool when we are not interested in exact estimation of the battery SOH, but we want to bracket the battery as new or old. On the other hand, when we want to estimate an accurate value of the SOH, we can use regression.

Similar to the earlier chapters, our focus in this chapter is on developing inner understanding of the machine learning algorithms. As such, we will focus on developing a generic framework for machine learning and demonstrate that various popular machine learning approaches are algorithms for numerical implementation of this generic framework. Though we restrict our discussion for artificial neural networks, support vector machines, and Gaussian processes, the overall approach can be applied to other machine learning approaches including more advanced deep learning algorithms.

8.2 Primer on Machine Learning Algorithms

Let us assume that we are given a set of n-dimensional input vector $\boldsymbol{x} \in \mathscr{R}^n$. For each input vector $\boldsymbol{x}$, the corresponding d-dimensional output vector $\boldsymbol{y} \in \mathscr{R}^d$ is also made available. The available set of all these input–output pairs, denoted in this chapter by $\{\boldsymbol{X}, \boldsymbol{Y}\}$, is known as the training dataset. Given the training dataset, any (supervised) machine learning algorithm is concerned with finding an optimal input–output mapping. This mapping is represented using a parametric function as

$$\boldsymbol{y} = f(\boldsymbol{x}, \boldsymbol{\theta}), \tag{8.1}$$

where $\boldsymbol{\theta}$ are unknown parameters. Any machine learning algorithm can be posed as a solution to the problem of estimating the parameters $\boldsymbol{\theta}$ using the training dataset.

Following the approach developed in the last chapter, we pose this parameter estimation as a Bayesian inference given by

$$p(\boldsymbol{\theta}|\boldsymbol{X}, \boldsymbol{Y}) \propto p(\boldsymbol{Y}|\boldsymbol{X}, \boldsymbol{\theta})\, p(\boldsymbol{\theta}). \tag{8.2}$$

The posterior distribution $p(\boldsymbol{\theta}|\boldsymbol{X}, \boldsymbol{Y})$ provides complete probabilistic information. Though this probabilistic information is important, in this chapter, we will confine our discussion to the point estimate of the parameters $\boldsymbol{\theta}$. This point estimate is given by

$$\hat{\boldsymbol{\theta}} = \arg\max_{\boldsymbol{\theta}} p\left(\boldsymbol{\theta}|\boldsymbol{X}, \boldsymbol{Y}\right). \tag{8.3}$$

Solution of Eq. 8.3 is known as a Maximum a-posteriory (MAP) estimate. In Eq. 8.2, if we use a completely non-informative prior, i.e.,

$$p\left(\boldsymbol{\theta}\right) \propto 1, \tag{8.4}$$

we obtain

$$p\left(\boldsymbol{\theta}|\boldsymbol{X}, \boldsymbol{Y}\right) \propto p\left(\boldsymbol{Y}|\boldsymbol{X}, \boldsymbol{\theta}\right). \tag{8.5}$$

Thus for the completely non-informative prior, the MAP reduces to

$$\hat{\boldsymbol{\theta}} = \arg\max_{\boldsymbol{\theta}} p\left(\boldsymbol{Y}|\boldsymbol{\theta}, \boldsymbol{X}\right). \tag{8.6}$$

Solution of Eq. 8.6 is known as a Maximum Likelihood Estimate (MLE) of $\boldsymbol{\theta}$. State-of-the-art machine learning algorithms are primarily concerned with this MLE estimate of $\boldsymbol{\theta}$. This generic machine learning approach can be implemented using Algorithm 5.

Algorithm 5 Machine Learning Algorithm

1: Obtain a dataset $\{\boldsymbol{X}, \boldsymbol{Y}\}$
2: Divide the dataset into training data $\{\boldsymbol{X}_T, \boldsymbol{Y}_T\}$ and testing data $\{\boldsymbol{X}_S, \boldsymbol{Y}_S\}$
3: Define the likelihood function $p\left(\boldsymbol{Y}_T|\boldsymbol{\theta}, \boldsymbol{X}_T\right)$
4: Implement an appropriate optimization method to obtain the MLE $\hat{\boldsymbol{\theta}} = \arg\max_{\boldsymbol{\theta}} p\left(\boldsymbol{Y}_T|\boldsymbol{\theta}, \boldsymbol{X}_T\right)$
5: $\hat{\boldsymbol{\theta}}$ can be used on the testing data for prediction as $\hat{\boldsymbol{Y}}_S = f\left(\boldsymbol{X}_S, \hat{\boldsymbol{\theta}}\right)$.

Depending on the nature of output, supervised learning problem is posed either as a regression or a classification problem. When the output is continuous, the supervised machine learning problem is known as regression and when the output is discrete, it is known as classification. As we demonstrate next, the machine learning approach presented in Algorithm 5 can be used for both regression and classification just by using an appropriate likelihood function.

8.2.1 Machine Learning for Regression

Let us consider a training dataset of size N, i.e., $\boldsymbol{X} = \{\boldsymbol{x}_i; \quad i = 1, ..., N\}$ and $\boldsymbol{Y} = \{\boldsymbol{y}_i; \quad i = 1, ..., N\}$.[1] We assume a parametric form given by Eq. 8.1 to define the input–output mapping on this training dataset. We further assume that the output

[1] For notational convenience, we have removed subscripts T and S for denoting the training and testing datasets. However, the meaning should be clear from the context.

vectors $\boldsymbol{y}_i$ are i.i.d. samples from a normal distribution

$$p\left(\boldsymbol{y}_i|\boldsymbol{\theta},\boldsymbol{x}_i\right) = \frac{1}{\sqrt{2\pi}} \exp\left(-\frac{1}{2}\left(\boldsymbol{y}_i - f\left(\boldsymbol{x}_i,\boldsymbol{\theta}\right)\right)^2\right). \tag{8.7}$$

Using these choices, the likelihood function is given by

$$p\left(\boldsymbol{Y}|\boldsymbol{\theta},\boldsymbol{X}\right) = \prod_{i=1}^{N} \frac{1}{\sqrt{2\pi}} \exp\left(-\frac{1}{2}\left(\boldsymbol{y}_i - f\left(\boldsymbol{x}_i,\boldsymbol{\theta}\right)\right)^2\right). \tag{8.8}$$

We can estimate the parameters $\boldsymbol{\theta}$ by maximizing the likelihood function (8.8). However, due to numerical difficulties, maximization of the likelihood function is seldom used in the machine learning algorithms. An alternate objective function, known as a log-likelihood function, provides a better numerical tractability for the machine learning algorithms. To obtain the log-likelihood function, take a natural logarithm of the likelihood function (8.8) as

$$\begin{aligned} \log\left(p\left(\boldsymbol{Y}|\boldsymbol{\theta},\boldsymbol{X}\right)\right) &= \log\left(\prod_{i=1}^{N} \frac{1}{\sqrt{2\pi}} \exp\left(-\frac{1}{2}\left(\boldsymbol{y}_i - f\left(\boldsymbol{x}_i,\boldsymbol{\theta}\right)\right)^2\right)\right) \\ &= -\frac{N}{2}\log(2\pi) - \frac{1}{2}\sum_{i=1}^{N}\left(\boldsymbol{y}_i - f\left(\boldsymbol{x}_i,\boldsymbol{\theta}\right)\right)^2. \end{aligned} \tag{8.9}$$

Since the constant term $-\frac{N}{2}\log(2\pi)$ in Eq. 8.9 does not affect the optimum, we will neglect this term for parameter estimation formulation. We also take negative of the log-likelihood function to convert the maximization problem to the minimization. Using these changes, we define the negative log-likelihood function as

$$\mathscr{L}\left(\boldsymbol{\theta}\right) = \frac{1}{2}\sum_{i=1}^{N}\left(\boldsymbol{y}_i - f\left(\boldsymbol{x}_i,\boldsymbol{\theta}\right)\right)^2. \tag{8.10}$$

Thus the maximum likelihood estimate of the parameters $\boldsymbol{\theta}$ is given by

$$\hat{\boldsymbol{\theta}} = \arg\min_{\boldsymbol{\theta}}\left[\frac{1}{2}\sum_{i=1}^{N}\left(\boldsymbol{y}_i - f\left(\boldsymbol{x}_i,\boldsymbol{\theta}\right)\right)^2\right]. \tag{8.11}$$

Note that the negative log-likelihood defined in Eq. 8.10 is same as the sum of squared error in the prediction $f\left(\boldsymbol{x}_i,\boldsymbol{\theta}\right)$. Thus the MLE of $\boldsymbol{\theta}$ obtained using Eq. 8.11 is equivalent to the minimization of squared error in predictions.

8.2.2 Machine Learning for Classification

In the classification problem, we are interested in assigning the input vector $\boldsymbol{x}_i$ to one of the classes in $\boldsymbol{C} = \{\mathscr{C}_1, \mathscr{C}_2, ..., \mathscr{C}_d\}$. Machine learning algorithms often focus on the probabilistic classification, where the aim is to obtain probability of the input vector belonging to one of the classes. Training data for the classification consists of a set of input vectors $\boldsymbol{x}_i$ and the associated class labels. In the typical dataset, one of the classes will have label 1 and all the other classes will be labeled as 0.

Similar to the regression, we define the likelihood function by assuming that each datapoint is independent and identically distributed. Using independence of the samples, the likelihood function is given by

$$p\left(\boldsymbol{Y}|\boldsymbol{\theta}, \boldsymbol{X}\right) = \prod_{i=1}^{N} p\left(\boldsymbol{y}_i|\boldsymbol{\theta}, \boldsymbol{x}_i\right). \tag{8.12}$$

For a given input vector $\boldsymbol{x}_i$, only one of the classes is active. Further, the classes are independent of each other. We use these properties to define $p\left(\boldsymbol{y}_i|\boldsymbol{\theta}, \boldsymbol{x}_i\right)$ as

$$p\left(\boldsymbol{y}_i|\boldsymbol{\theta}, \boldsymbol{x}_i\right) = \prod_{j=1}^{d} p\left(y_i^j|\boldsymbol{\theta}, \boldsymbol{x}_i\right)^{y_i^j}, \tag{8.13}$$

where y_i^j is jth component of $\boldsymbol{y}_i$. Thus,

$$p\left(\boldsymbol{Y}|\boldsymbol{\theta}, \boldsymbol{X}\right) = \prod_{i=1}^{N} \prod_{j=1}^{d} p\left(y_i^j|\boldsymbol{\theta}, \boldsymbol{x}_i\right)^{y_i^j}. \tag{8.14}$$

Using the procedure outlined for regression, the negative log-likelihood function for the classification is given by

$$\mathscr{L}\left(\boldsymbol{\theta}\right) = -\sum_{i=1}^{N} \sum_{j=1}^{d} y_i^j \log\left(p\left(y_i^j|\boldsymbol{\theta}, \boldsymbol{x}_i\right)\right). \tag{8.15}$$

This negative log-likelihood function is known as the cross-entropy.

Definition of the negative log-likelihood is completed by specifying a form for the probability distribution $p\left(y_i^j|\boldsymbol{\theta}, \boldsymbol{x}_i\right)$. For classification, we use the following probability distribution for $p\left(y_i^j|\boldsymbol{\theta}, \boldsymbol{x}_i\right)$:

$$p\left(y_i^j|\boldsymbol{\theta}, \boldsymbol{x}_i\right) = \frac{\exp\left(f_j\left(\boldsymbol{x}_i, \boldsymbol{\theta}\right)\right)}{\sum_{k=1}^{d} \exp\left(f_k\left(\boldsymbol{x}_i, \boldsymbol{\theta}\right)\right)}, \tag{8.16}$$

where $f_j(\cdot, \cdot)$ denotes jth component of the prediction. Probability distribution specified in Eq. 8.16 is known as the softmax distribution. Using Eq. 8.16 with Eq. 8.15, the negative log-likelihood is given by

$$\mathscr{L}(\boldsymbol{\theta}) = -\sum_{i=1}^{N}\sum_{j=1}^{d} y_i^j \log\left(\frac{\exp\left(f_j\left(\boldsymbol{x}_i, \boldsymbol{\theta}\right)\right)}{\sum_{k=1}^{d}\exp\left(f_k\left(\boldsymbol{x}_i, \boldsymbol{\theta}\right)\right)}\right). \tag{8.17}$$

Using the softmax distribution, MLE estimate of the parameters is given by

$$\hat{\boldsymbol{\theta}} = \arg\min_{\boldsymbol{\theta}}\left[-\sum_{i=1}^{N}\sum_{j=1}^{d} y_i^j \log\left(\frac{\exp\left(f_j\left(\boldsymbol{x}_i, \boldsymbol{\theta}\right)\right)}{\sum_{k=1}^{d}\exp\left(f_k\left(\boldsymbol{x}_i, \boldsymbol{\theta}\right)\right)}\right)\right]. \tag{8.18}$$

8.2.3 Gradient Descent Algorithm

In this section, we have presented a generic framework for machine learning algorithms and demonstrated how regression and classification problems can be derived from this generic framework. We follow the same premise in this subsection and describe a generic algorithm for solution of the machine learning problems. In particular, we will describe the gradient descent algorithm, which is one of the most widely used algorithms for both regression and classification.

Objective of the machine learning algorithm is to minimize the negative log-likelihood function $\mathscr{L}(\boldsymbol{\theta})$. We can visualize $\mathscr{L}(\boldsymbol{\theta})$ as a terrain consisting of troughs and peaks. The local minima are located at the bottom-most point of the nearest trough. From any position, we can reach this bottom-most point fastest by sliding on the slope. This slope is given by the gradient $\nabla_{\boldsymbol{\theta}}\mathscr{L}(\boldsymbol{\theta})$. As we descent along the gradient, the resultant algorithm is known as the gradient descent algorithm. This algorithm is implemented iteratively as follows. If the parameter values for the current iteration are $\boldsymbol{\theta}_i$, then the values at the next iteration are given by

$$\boldsymbol{\theta}_{i+1} = \boldsymbol{\theta}_i - \alpha\nabla_{\boldsymbol{\theta}}\mathscr{L}(\boldsymbol{\theta}), \tag{8.19}$$

where α is known as learning rate. In the machine learning community, each iteration of the gradient descent algorithm is known as *epoch*.

Though easy to implement, gradient descent formulation given by Eq. 8.19 often oscillates near the local optima. Convergence of the gradient descent algorithm is slow due to these oscillations. One method to dampen the oscillations of the gradient descent algorithm is to use the concept of momentum. Slide along the slope results in the momentum gain, which in turn increases velocity of the slide. To implement this concept, we define a velocity vector $\boldsymbol{v}$ of the size same as $\boldsymbol{\theta}$. This velocity vector is updated as

$$v_{i+1} = \gamma v_i + \alpha \nabla_{\theta} \mathscr{L}(\theta), \tag{8.20}$$

where $\gamma \in [0, 1)$ determines the number of previous gradients incorporated in the current update. Using the updated velocity vector, the parameter update is obtained as

$$\theta_{i+1} = \theta_i - v_{i+1}. \tag{8.21}$$

The gradient $\nabla_{\theta} \mathscr{L}(\theta)$ is defined for a complete dataset. However, for a very large dataset, evaluation of the gradient is computationally intractable. Hence, a variant of the gradient descent algorithm, known as mini-batch stochastic gradient descent algorithm, is used in practice. In this algorithm, a small subset of the complete dataset is selected at random. This small subset is known as a mini-batch. For a given mini-batch, the gradient descent algorithm is used to obtain local minima. Next mini-batch is then selected randomly and the gradient descent algorithm is restarted from the already obtained local minima. This process is continued iteratively for a user-defined number of maximum iterations. The mini-batch stochastic gradient descent algorithm is summarized in Algorithm (6).

Algorithm 6 Mini-batch Stochastic Gradient Descent Algorithm

Obtain a training dataset $\{X, Y\}$.
Define a parametric form for the function $y = f(x, \theta)$.
Define a likelihood function $p(Y|\theta, X)$. For the likelihood function, obtain an analytical form for the negative log-likelihood $\mathscr{L}(\theta) = -\log[p(Y|\theta, X)]$.
Specify batch size, number of batches, number of epochs, initial guess for parameters θ_0, learning rate α, momentum γ, and velocity vector v.
for `i=1 TO number_of_epochs` **do**
  for `j=0 TO number_of_batches` **do**
    Randomly select a mini-batch $\{X_j, Y_j\}$ of size `batch_size`.
    Using the mini-batch $\{X_i, Y_i\}$, calculate the gradient $\nabla_{\theta} \mathscr{L}(\theta)$
    Update the velocity vector as

$$v_{j+1} = \gamma v_j + \alpha \nabla_{\theta} \mathscr{L}(\theta).$$

    Update the parameters as

$$\theta_{j+1} = \theta_j - v_{j+1}.$$

  end for
  Set $\theta_0 = \theta_j$
end for
Set $\hat{\theta} = \theta_i$.

Gradient evaluation is critical for an efficient implementation of Algorithm (6). We finish this section by deriving gradient for both regression and classification. For regression, gradient of the negative log-likelihood is given by

$$\nabla_{\theta}\mathcal{L}(\boldsymbol{\theta}) = \nabla_{\theta}\left[\frac{1}{2}\sum_{i=1}^{N}\left(\boldsymbol{y}_i - f(\boldsymbol{x}_i, \boldsymbol{\theta})\right)^2\right]$$
$$= -\left(\boldsymbol{y}_i - f(\boldsymbol{x}_i, \boldsymbol{\theta})\right)\nabla_{\theta} f(\boldsymbol{x}_i, \boldsymbol{\theta}). \tag{8.22}$$

Gradient for the classification can be obtained similarly; however, the derivation is more involved as compared to the regression. Gradient of the negative log-likelihood defined in Eq. 8.17 is given by

$$\nabla_{\theta}\mathcal{L}(\boldsymbol{\theta}) = \nabla_{\theta}\left[-\sum_{i=1}^{N}\sum_{j=1}^{d} y_i^j \log\left(p\left(y_i^j|\boldsymbol{\theta}, \boldsymbol{x}_i\right)\right)\right]. \tag{8.23}$$

When we use the softmax distribution (8.16) for $p\left(y_i^j|\boldsymbol{\theta}, \boldsymbol{x}_i\right)$, the negative log-likelihood becomes function of the mapping $f(\boldsymbol{x}_i, \boldsymbol{\theta})$. Using the chain rule, we obtain

$$\nabla_{\theta}\mathcal{L}(\boldsymbol{\theta}) = \sum_{p=1}^{d}\frac{\partial\mathcal{L}(\boldsymbol{\theta})}{\partial f_p(\boldsymbol{x}_i, \boldsymbol{\theta})}\nabla_{\theta} f_p(\boldsymbol{x}_i, \boldsymbol{\theta}). \tag{8.24}$$

Next, we consider evaluation of the partial derivative $\frac{\partial\mathcal{L}(\boldsymbol{\theta})}{\partial f_p(\boldsymbol{x}_i, \boldsymbol{\theta})}$. Using the negative log-likelihood defined in Eq. 8.17, the partial derivative is given by

$$\frac{\partial\mathcal{L}(\boldsymbol{\theta})}{\partial f_p(\boldsymbol{x}_i, \boldsymbol{\theta})} = -\sum_{i=1}^{N}\sum_{j=1}^{d} y_i^j \frac{\partial \log\left(p\left(y_i^j|\boldsymbol{\theta}, \boldsymbol{x}_i\right)\right)}{f_p(\boldsymbol{x}_i, \boldsymbol{\theta})}$$
$$= -\sum_{i=1}^{N}\sum_{j=1}^{d} y_i^j \frac{1}{p\left(y_i^j|\boldsymbol{\theta}, \boldsymbol{x}_i\right)}\frac{\partial\left(p\left(y_i^j|\boldsymbol{\theta}, \boldsymbol{x}_i\right)\right)}{\partial f_p(\boldsymbol{x}_i, \boldsymbol{\theta})}. \tag{8.25}$$

Using softmax distribution, the partial derivative is given by

$$\frac{\partial\left(p\left(y_i^j|\boldsymbol{\theta}, \boldsymbol{x}_i\right)\right)}{\partial f_p(\boldsymbol{x}_i, \boldsymbol{\theta})} = \frac{\partial}{\partial f_p(\boldsymbol{x}_i, \boldsymbol{\theta})}\left(\frac{\exp\left(f_j(\boldsymbol{x}_i, \boldsymbol{\theta})\right)}{\sum_{k=1}^{d}\exp\left(f_k(\boldsymbol{x}_i, \boldsymbol{\theta})\right)}\right). \tag{8.26}$$

If we use the quotient rule when $p = j$, we obtain

$$\frac{\partial\left(p\left(y_i^j|\boldsymbol{\theta},\boldsymbol{x}_i\right)\right)}{\partial f_p\left(\boldsymbol{x}_i,\boldsymbol{\theta}\right)} = \frac{\left[\sum_{k=1}^d \exp\left(f_k(\boldsymbol{x}_i,\boldsymbol{\theta})\right)\right]\exp\left(f_p(\boldsymbol{x}_i,\boldsymbol{\theta})\right) - \exp\left(f_p(\boldsymbol{x}_i,\boldsymbol{\theta})\right)\exp\left(f_p(\boldsymbol{x}_i,\boldsymbol{\theta})\right)}{\left[\sum_{k=1}^d \exp\left(f_k(\boldsymbol{x}_i,\boldsymbol{\theta})\right)\right]^2}$$

$$= \frac{\exp\left(f_p(\boldsymbol{x}_i,\boldsymbol{\theta})\right)}{\sum_{k=1}^d \exp\left(f_k(\boldsymbol{x}_i,\boldsymbol{\theta})\right)} - \frac{\exp\left(f_p(\boldsymbol{x}_i,\boldsymbol{\theta})\right)}{\sum_{k=1}^d \exp\left(f_k(\boldsymbol{x}_i,\boldsymbol{\theta})\right)}\frac{\exp\left(f_p(\boldsymbol{x}_i,\boldsymbol{\theta})\right)}{\sum_{k=1}^d \exp\left(f_k(\boldsymbol{x}_i,\boldsymbol{\theta})\right)}$$

$$= p\left(y_i^p|\boldsymbol{\theta},\boldsymbol{x}_i\right)\left(1 - p\left(y_i^p|\boldsymbol{\theta},\boldsymbol{x}_i\right)\right). \tag{8.27}$$

Similarly when $p \neq j$, we obtain

$$\frac{\partial\left(p\left(y_i^j|\boldsymbol{\theta},\boldsymbol{x}_i\right)\right)}{\partial f_p\left(\boldsymbol{x}_i,\boldsymbol{\theta}\right)} = -p\left(y_i^p|\boldsymbol{\theta},\boldsymbol{x}_i\right)p\left(y_i^j|\boldsymbol{\theta},\boldsymbol{x}_i\right). \tag{8.28}$$

Using the partial derivatives (8.27) and (8.28) in (8.28), we obtain

$$\frac{\partial\mathscr{L}\left(\boldsymbol{\theta}\right)}{\partial f_p\left(\boldsymbol{x}_i,\boldsymbol{\theta}\right)} = -\sum_{i=1}^N\left[y_i^p - p\left(y_i^p|\boldsymbol{\theta},\boldsymbol{x}_i\right)\sum_{j=1}^d y_i^j\right]. \tag{8.29}$$

Since only one of the classes is active, $\sum_{j=1}^d y_i^j = 1$. Thus,

$$\frac{\partial\mathscr{L}\left(\boldsymbol{\theta}\right)}{\partial f_p\left(\boldsymbol{x}_i,\boldsymbol{\theta}\right)} = -\sum_{i=1}^N\left[y_i^p - \frac{\exp\left(f_j\left(\boldsymbol{x}_i,\boldsymbol{\theta}\right)\right)}{\sum_{k=1}^d \exp\left(f_k\left(\boldsymbol{x}_i,\boldsymbol{\theta}\right)\right)}\right]. \tag{8.30}$$

Thus, gradient of the negative log-likelihood for a classification problem is given by

$$\nabla_{\boldsymbol{\theta}}\mathscr{L}\left(\boldsymbol{\theta}\right) = -\sum_{i=1}^N\sum_{p=1}^d\left[y_i^p - \frac{\exp\left(f_p\left(\boldsymbol{x}_i,\boldsymbol{\theta}\right)\right)}{\sum_{k=1}^d \exp\left(f_k\left(\boldsymbol{x}_i,\boldsymbol{\theta}\right)\right)}\right]\nabla_{\boldsymbol{\theta}} f_p\left(\boldsymbol{x}_i,\boldsymbol{\theta}\right). \tag{8.31}$$

From the gradient formulations obtained for regression and classification, we can make following two observations:

- Form of $\nabla_{\boldsymbol{\theta}}\mathscr{L}\left(\boldsymbol{\theta}\right)$ is similar for both the regression and classification problems. Thus, a similar algorithm can be used for minimizing $\mathscr{L}\left(\boldsymbol{\theta}\right)$.
- For a given parametric form of the input–output mapping, $\nabla_{\boldsymbol{\theta}} f\left(\cdot,\boldsymbol{\theta}\right)$ is the only unknown term of the gradient.

We can obtain a given machine learning algorithm by specifying a particular parametric form for this input–output mapping. In the remaining of this chapter, we will review some popular machine learning algorithms commonly used by the Li-ion battery community.

8.3 Artificial Neural Network

As the name suggests, an artificial neural network (ANN) is a graphical model that mimics working of biological neurons in a brain. A typical neuron in the ANN is shown in Fig. 8.1.

Neuron is a computational unit that transforms the input vector $\boldsymbol{x}$ using two simple mathematical operations. In the first operation, the input vector is linearly transformed as

$$z = \boldsymbol{W}\boldsymbol{x} + b, \tag{8.32}$$

where $\boldsymbol{W}$ is a weight vector and b is known as bias. In the second operation, a nonlinear function is applied on z to obtain output of the neuron. This nonlinear function is known as an *activation function*. By using different activation functions, we can obtain different flavors of ANN. Some of the most commonly used activation functions are sigmoid, hyperbolic tan (tanh), and rectified linear function. In this chapter, we will focus on the sigmoid function. However, the discussion is generic in nature and can be used for other activation functions also. The sigmoid function is given by

$$\sigma(z) = \frac{1}{1 + \exp(-z)}. \tag{8.33}$$

For future use, we derive here the derivative of this sigmoid function as follows:

$$\begin{aligned}
\frac{d\sigma(z)}{dz} &= \frac{d}{dz}\left(\frac{1}{1+\exp(-z)}\right) \\
&= \frac{\exp(-z)}{[1+\exp(-z)]^2} \\
&= \frac{1}{1+\exp(-z)}\left[1 - \frac{1}{1+\exp(-z)}\right] \\
&= \sigma(z)(1-\sigma(z)).
\end{aligned} \tag{8.34}$$

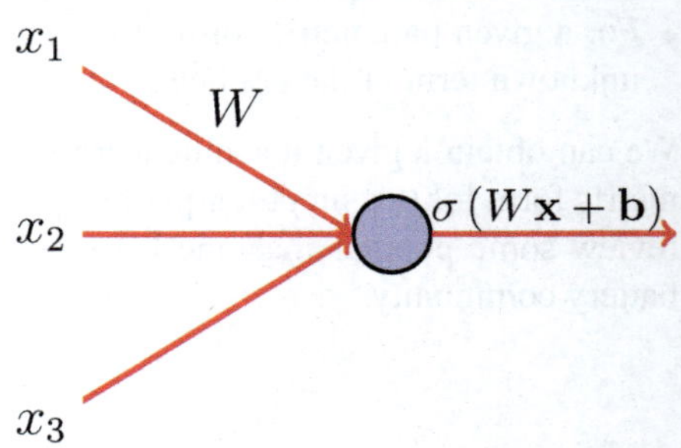

Fig. 8.1 A typical neuron

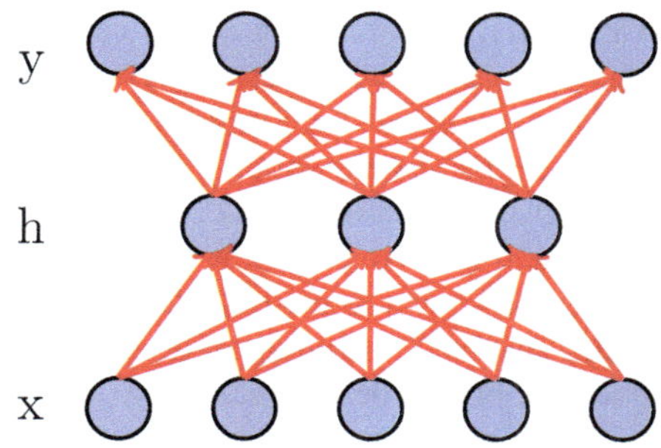

Fig. 8.2 Three-layer artificial neural network

A neural network is constructed by combining these neurons together in the form of layers. A typical artificial neural network is shown in Fig. 8.2. Each layer of the neural network consists of predetermined number of nodes. Bottom-most layer is known as the input layer. Number of nodes in this input layer is equal to the dimension of the input vector $\boldsymbol{x}$. Note that the nodes in the input layer are not neurons. The topmost layer is known as the output layer with the number of nodes given by the dimension of the output vector $\boldsymbol{y}$. The input and output layers are known as the visible layers.

If we directly connect the input and output layers with the sigmoid activation function, we obtain a simplest neural network, known as the *logistic regression* model. To define this logistic regression model, we first obtain a linear mapping

$$z = \boldsymbol{W}\boldsymbol{x} + \boldsymbol{b}. \tag{8.35}$$

Next, we apply sigmoid activation function to obtain a parametric input–output mapping form of the logistic regression model as

$$f(\boldsymbol{x}, \boldsymbol{\theta}) = \sigma(z) = \frac{1}{1 + \exp(-(z))}. \tag{8.36}$$

Parameters of this model are given by $\boldsymbol{\theta} = [\boldsymbol{W}, \boldsymbol{b}]$. Thus the gradient is given by

$$\nabla_{\boldsymbol{\theta}} f(\boldsymbol{x}, \boldsymbol{\theta}) = \left[\frac{\partial \sigma(\boldsymbol{W}\boldsymbol{x} + \boldsymbol{b})}{\partial \boldsymbol{W}}; \quad \frac{\partial \sigma(\boldsymbol{W}\boldsymbol{x} + \boldsymbol{b})}{\partial \boldsymbol{b}}\right]. \tag{8.37}$$

Using chain rule with derivative of the sigmoid function derived earlier, the partial derivatives are given by

$$\begin{aligned}
\frac{\partial \sigma(\boldsymbol{W}\boldsymbol{x} + \boldsymbol{b})}{\partial \boldsymbol{W}} &= \sigma(\boldsymbol{W}\boldsymbol{x} + \boldsymbol{b})(1 - \sigma(\boldsymbol{W}\boldsymbol{x} + \boldsymbol{b}))\boldsymbol{x} \\
\frac{\partial \sigma(\boldsymbol{W}\boldsymbol{x} + \boldsymbol{b})}{\partial \boldsymbol{b}} &= \sigma(\boldsymbol{W}\boldsymbol{x} + \boldsymbol{b})(1 - \sigma(\boldsymbol{W}\boldsymbol{x} + \boldsymbol{b})).
\end{aligned} \tag{8.38}$$

The partial derivatives in (8.38) can be used with the stochastic gradient descent algorithm to train the logistic regression model.

Though simple, the logistic regression model is rigid as structure of the network is determined by the input–output vectors. Due to this rigid structure, accuracy of the logistic regression model is often low. We can improve accuracy of the neural network by introducing new layers between the input–output layers. These middle layers are known as the hidden layers. Unlike the visible layers where a number of nodes are fixed, a number of nodes of the hidden layer are user-defined and can be fixed to obtain a desired accuracy.

To define a parametric input–output mapping form for the ANN, we first define a linear mapping from the input to the hidden layer as

$$z_h = \boldsymbol{W}_1 \boldsymbol{x} + \boldsymbol{b}_1, \tag{8.39}$$

where $\boldsymbol{W}_1$ and $\boldsymbol{b}_1$ are weight and bias matrices with appropriate dimensions. Sigmoid activation function is applied on this linear mapping to obtain the output of the hidden layer as

$$\boldsymbol{h} = \frac{1}{1 + \exp\left(-\left(z_h\right)\right)}. \tag{8.40}$$

This hidden layer output is supplied as an input to the topmost layer. Again, we first define the linear mapping as

$$z_y = \boldsymbol{W}_2 \boldsymbol{h} + \boldsymbol{b}_2. \tag{8.41}$$

Then the topmost layer output is given by

$$\hat{y} = \frac{1}{1 + \exp\left(-\left(z_y\right)\right)}. \tag{8.42}$$

If we represent the neural network defined using Eqs. 8.40 and 8.42 in a parametric input–output mapping form, we can apply the stochastic gradient descent algorithm. For our three-layer neural network considered in this chapter, the input–output mapping is given by

$$f\left(\boldsymbol{x}, \boldsymbol{\theta}\right) = \sigma\left(\boldsymbol{W}_2 \sigma\left(\boldsymbol{W}_1 \boldsymbol{x} + \boldsymbol{b}_1\right) + \boldsymbol{b}_2\right). \tag{8.43}$$

Note that function (8.43) has a nested hierarchical structure. Although we use a simple activation function, this nested hierarchical structure allows approximation of complex functions with high accuracy.

Parameters of this three-layer neural network are given by $\boldsymbol{\theta} = [\boldsymbol{W}_1, \boldsymbol{b}_1, \boldsymbol{W}_2, \boldsymbol{b}_1]$. We can use this neural network for prediction only after estimating the parameters $\boldsymbol{\theta}$, i.e., we have to first train the neural network. We describe this training procedure in the following. Let $\{\boldsymbol{X}, \boldsymbol{Y}\}$ be our training dataset. From this complete dataset, we can select a mini-batch $\{\boldsymbol{X}_i, \boldsymbol{Y}_i\}$ at random.

For inputs $\boldsymbol{X}_i$, the output of the hidden layer is given by

$$\boldsymbol{Z}_h = \boldsymbol{W}_1\boldsymbol{X}_i + \boldsymbol{b}_1 \tag{8.44}$$

$$\boldsymbol{H}_i = \sigma\left(\boldsymbol{Z}_h\right). \tag{8.45}$$

Using the hidden layer output, prediction of $\boldsymbol{Y}_i$ is given by

$$\boldsymbol{Z}_y = \boldsymbol{W}_2\boldsymbol{H}_i + \boldsymbol{b}_2 \tag{8.46}$$

$$\hat{\boldsymbol{Y}}_i = \sigma\left(\boldsymbol{Z}_y\right). \tag{8.47}$$

Thus for a given set of parameters, we can make one forward pass to obtain predictions $\hat{\boldsymbol{Y}}_i$ from the input $\boldsymbol{X}_i$. To train our neural network, first we define the error in this prediction. For a regression problem, the error is given by

$$\mathscr{E} = \left(\boldsymbol{Y}_i - \hat{\boldsymbol{Y}}_i\right). \tag{8.48}$$

Similarly, the error for a classification problem is

$$\mathscr{E} = \left(\boldsymbol{Y}_i - \frac{\exp\left(\hat{\boldsymbol{Y}}_i\right)}{\sum_{j=1}^{d} \exp\left(\hat{Y}_j^i\right)}\right). \tag{8.49}$$

We further need gradient of the function (8.43) for training the neural network. Using the chain rule and the derivative of the sigmoid activation function, partial derivative w.r.t. $\boldsymbol{W}_2$ is given by

$$\begin{aligned}\frac{\partial \hat{\boldsymbol{Y}}_i}{\partial \boldsymbol{W}_2} &= \frac{\partial}{\partial \boldsymbol{Z}_h}\left(\sigma\left(\boldsymbol{Z}_h\right)\right)\frac{\partial \boldsymbol{Z}_h}{\partial \boldsymbol{W}_2}\\ &= \hat{\boldsymbol{Y}}_i\left(1 - \hat{\boldsymbol{Y}}_i\right)\boldsymbol{H}_i.\end{aligned} \tag{8.50}$$

Similarly, the partial derivative w.r.t $\boldsymbol{b}_2$ is given by

$$\begin{aligned}\frac{\partial \hat{\boldsymbol{Y}}_i}{\partial \boldsymbol{b}_2} &= \frac{\partial}{\partial \boldsymbol{Z}_h}\left(\sigma\left(\boldsymbol{Z}_h\right)\right)\frac{\partial \boldsymbol{Z}_h}{\partial \boldsymbol{b}_2}\\ &= \hat{\boldsymbol{Y}}_i\left(1 - \hat{\boldsymbol{Y}}_i\right).\end{aligned} \tag{8.51}$$

We can follow the similar procedure to obtain partial derivatives w.r.t. $\boldsymbol{W}_1$ and $\boldsymbol{b}_1$ as follows:

$$\frac{\partial \hat{\boldsymbol{Y}}_i}{\partial \boldsymbol{W}_1} = \frac{\partial}{\partial \boldsymbol{Z}_y}\left(\sigma\left(\boldsymbol{Z}_y\right)\right)\frac{\partial \boldsymbol{Z}_y}{\partial \boldsymbol{H}_i}\frac{\partial \boldsymbol{H}_i}{\partial \boldsymbol{Z}_h}\frac{\partial \boldsymbol{Z}_h}{\partial \boldsymbol{W}_1}$$
$$= \hat{\boldsymbol{Y}}_i\left(1-\hat{\boldsymbol{Y}}_i\right)\boldsymbol{W}_2\boldsymbol{H}_i\left(1-\boldsymbol{H}_i\right)\boldsymbol{X}_i. \tag{8.52}$$

$$\frac{\partial \hat{\boldsymbol{Y}}_i}{\partial \boldsymbol{b}_1} = \frac{\partial}{\partial \boldsymbol{Z}_y}\left(\sigma\left(\boldsymbol{Z}_y\right)\right)\frac{\partial \boldsymbol{Z}_y}{\partial \boldsymbol{H}_i}\frac{\partial \boldsymbol{H}_i}{\partial \boldsymbol{Z}_h}\frac{\partial \boldsymbol{Z}_h}{\partial \boldsymbol{b}_1}$$
$$= \hat{\boldsymbol{Y}}_i\left(1-\hat{\boldsymbol{Y}}_i\right)\boldsymbol{W}_2\boldsymbol{H}_i\left(1-\boldsymbol{H}_i\right). \tag{8.53}$$

Comparing Eqs. 8.50–8.51 with 8.52–8.53, we note that the output of upper layers is propagated backward to the lower layers. Thus, the resultant algorithm is known as the *backpropagation algorithm*. Training procedure for the neural network is summarized in Algorithm (7).

Algorithm 7 Mini-batch Training of Artificial Neural Network

Obtain a training dataset $\{\boldsymbol{X}, \boldsymbol{Y}\}$.
Specify number of hidden layers and number of nodes in each hidden layer.
Select the activation function $\sigma(\cdot)$.
Specify the loss function depending on type of the problem (regression or classification).
Specify batch size, number of batches, number of epochs, initial guess for parameters $\boldsymbol{\theta}_0$, learning rate α, and momentum γ.
for `i=1 TO number_of_epochs` **do**
    for `j=0 TO number_of_batches` **do**
        Randomly select a mini-batch $\{\boldsymbol{X}_j, \boldsymbol{Y}_j\}$ of size `batch_size`.
        Make one forward pass using the mini-batch.
        Use the backpropagation (Eqs. 8.50–8.53) to obtain the gradient $\nabla_{\boldsymbol{\theta}}\mathscr{L}(\boldsymbol{\theta})$
        Update the velocity vector as

$$\boldsymbol{v}_{j+1} = \gamma \boldsymbol{v}_j + \alpha \nabla_{\boldsymbol{\theta}}\mathscr{L}(\boldsymbol{\theta}).$$

        Update the parameters as

$$\boldsymbol{\theta}_{j+1} = \boldsymbol{\theta}_j - \boldsymbol{v}_{j+1}.$$

    end for
    Set $\boldsymbol{\theta}_0 = \boldsymbol{\theta}_j$
end for
Set $\hat{\boldsymbol{\theta}} = \boldsymbol{\theta}_j$.

8.4 Support Vector Machine

This section briefly introduces the Support Vector Machines (SVM) for classification and regression [56]. Without the loss of generality, the algorithm is presented for a two-class classification problem and subsequently the pointers are provided for its extension to the general classification and regression. Consider a two-class classification problem with a training dataset $(\mathbf{x}_i, \mathbf{y}_i; i = 1, \ldots, n)$, where $\mathbf{x}_i \in \mathbb{R}^n$

and $\mathbf{y} \in \{-1, 1\}$. The data point can be separated using a hyperplane given by

$$\langle \omega, \mathbf{x} \rangle + b = 0, \tag{8.54}$$

where ω is a parameter vector and $\langle \cdot, \cdot \rangle$ denotes a dot product. The classification problem is to find an optimal separating hyperplane, which maximizes the distance between itself and the nearest data point of each class. The optimal separating hyperplane must satisfy the canonical constraint

$$\mathbf{y}_i \left[\langle \omega, \mathbf{x}_i \rangle + b \right] = 1 - \varepsilon_i, \tag{8.55}$$

where ε_i is a non-negative measure of the misclassification error. The optimal separating hyperplane is defined using

$$\Phi(\omega, \xi) = \frac{1}{2} \|\omega\|^2 + C \sum_i \xi_i, \tag{8.56}$$

subject to the constraint in Eq. 8.55. Here, C is a user-defined value. The optimization problem is solved using Lagrangian multipliers, where the optimum is a saddle point of the Lagrangian

$$\Phi(\omega, b, \xi, \alpha, \beta) = \frac{1}{2} \|\omega\|^2 + C \sum_i \xi_i - \sum_i \alpha_i (\mathbf{y}_i [\omega^T \mathbf{x}_i + b] - 1 + \xi_i) - \sum_i \beta_i \xi_i, \tag{8.57}$$

while α and β are the Lagrange multipliers. The Lagrange multipliers are zero except for a small subset of the input vectors. This small subset is known as the support vectors (SVs) [56]. The SVM can be adapted for a regression problem using the SVR algorithm, where objective is to find an optimal function

$$f(\mathbf{x}) = \langle \omega, \mathbf{x} \rangle + b. \tag{8.58}$$

The goal of the SVR is to find a function (Eq. 8.58) such that the maximum deviation of $f(\mathbf{x})$ from an arbitrary training data is less than a user-defined value ε, while maintaining the highest possible flatness. The resultant optimization problem has a form

$$\Phi(\omega, \xi) = \frac{1}{2} \|\omega\|^2 + C \sum_i \xi_i \tag{8.59}$$

subject to the constraints

$$\mathbf{y}_i - \langle \omega, \mathbf{x}_i \rangle - b \le \varepsilon + \xi_i. \tag{8.60}$$

Required function is the optimal of the Lagrangian

$$\Phi(\omega, b, \xi, \alpha, \beta) = \frac{1}{2}\|\omega\|^2 + C\sum_i \xi_i - \sum_i \alpha_i([\omega^T \mathbf{x}_i + b] - \mathbf{y}_i + \varepsilon + \xi_i) - \sum_i \beta_i \xi_i. \quad (8.61)$$

The resultant optimized target function has the form

$$f(\mathbf{x}) = \sum_{i=1}^{L} \alpha_i \langle \mathbf{x}_i, \mathbf{x} \rangle + b, \quad (8.62)$$

where $\mathbf{x}_i$ are the support vectors.

The main advantage of the SVM algorithm for classification and regression is the final formulation in terms of the support vectors, which condenses the large training data to a significantly smaller subspace of SVs. Moreover, the formulation does not require any computationally intensive mathematical operations. The proposed method exploits these advantages of the SVM to obtain a computationally efficient RUL estimation algorithm.

8.5 Gaussian Process Emulator

In this chapter, we have posed a supervised learning problem from a parameter estimation perspective. For this, we first define a parametric function between the input $\boldsymbol{x}$ and the output $\boldsymbol{y}$ as

$$\boldsymbol{y} = f(\boldsymbol{x}, \boldsymbol{\theta}), \quad (8.63)$$

and subsequently estimate the parameters $\boldsymbol{\theta}$ using MLE. We have further investigated two different forms for the function $f(\cdot, \cdot)$, which resulted in the artificial neural network and the support vector machine algorithms. This approach has one drawback as accuracy of the predictions depends on the form of the function used. We can overcome this problem to a certain extent by using a more flexible and complex function; however, this often results in overfitting.

In this section, we will go a step further and investigate a scenario when the form for $f(\cdot, \cdot)$ is not specified. In this formulation, we make inference directly on the functions rather than the parameters. For this inference, we again use the Bayes theorem as

$$p(f|\boldsymbol{X}, \boldsymbol{Y}) \propto p(\boldsymbol{Y}|\boldsymbol{X}, f)\, p(f). \quad (8.64)$$

We can note one key difference between the formulations (8.2) and (8.64). In (8.2), we specified a probability distribution on the parameters $\boldsymbol{\theta}$, whereas in (8.64) we directly specify a probability distribution on the function $f(\cdot, \cdot)$. The formulation (8.64) implies that we first define a set of all the possible functions mapping $\boldsymbol{x}$ to

$\boldsymbol{y}$ and then specify a probability distribution on this set. However, definition of all the possible functions mapping $\boldsymbol{x}$ to $\boldsymbol{y}$ seems intractable. This seemingly intractable problem is resolved by using concept of the Gaussian process.

In essence, Gaussian process is a generalization of the multivariate Gaussian distribution. A multivariate Gaussian distribution is a probability distribution defined on a random vector. Similarly, we can consider a Gaussian process as a probability distribution defined on a random function. Similar to a random vector sampled from multivariate Gaussian distribution, we can sample a random function f from a Gaussian process. We formally define a Gaussian process by

$$f \sim GP\left(\boldsymbol{\mu}(\boldsymbol{x}); \mathscr{K}\left(\boldsymbol{x}, \boldsymbol{x}'\right)\right), \tag{8.65}$$

where $\boldsymbol{\mu}(\boldsymbol{x})$ is mean and $\mathscr{K}\left(\boldsymbol{x}, \boldsymbol{x}'\right)$ is a covariance function.

A Gaussian process defined in Eq. 8.65 allows us to specify a probability distribution on random functions. However, the function itself is defined at all the possible values of $\boldsymbol{x}$.[2] Thus to complete the definition, we need to specify the function output at all the possible values of $\boldsymbol{x}$. This essentially means we require infinitely many training data points. Next, we briefly describe how we can overcome this intractability.

Let us consider two sets of distinct input values $A = [\boldsymbol{x}_{i1};\quad i1 = 1, ..., n1]$ and $B = [\boldsymbol{x}_{i2};\quad i2 = 1, ..., n2]$. Let $\boldsymbol{f}_A$ and $\boldsymbol{f}_B$ denote the corresponding function evaluations at A and B. We define a multivariate Gaussian distribution on $\left[\boldsymbol{f}_A; \boldsymbol{f}_B\right]$ as

$$\begin{bmatrix} \boldsymbol{f}_A \\ \boldsymbol{f}_B \end{bmatrix} \sim \mathscr{N}\left(\begin{bmatrix} \boldsymbol{\mu}_A \\ \boldsymbol{\mu}_B \end{bmatrix}; \boldsymbol{K}\right), \tag{8.66}$$

where $\begin{bmatrix} \boldsymbol{\mu}_A \\ \boldsymbol{\mu}_B \end{bmatrix}$ is mean and $\boldsymbol{K}$ is a covariance matrix defined as

$$\boldsymbol{K} = \begin{bmatrix} \boldsymbol{K}_{AA} & \boldsymbol{K}_{AB} \\ \boldsymbol{K}_{BA} & \boldsymbol{K}_{BB} \end{bmatrix}. \tag{8.67}$$

In Eq. 8.67, $\boldsymbol{K}_{AA}$ is a covariance matrix defined on A, $\boldsymbol{K}_{BB}$ is a covariance matrix defined on B, and $\boldsymbol{K}_{AB}$ is a cross-covariance matrix between A and B. Following two properties of this multivariate Gaussian distribution are critical for us:

1. Marginals of a multivariate Gaussian distribution are also Gaussian. These marginal distributions are given by

$$\boldsymbol{f}_A \sim \mathscr{N}\left(\boldsymbol{\mu}_A; \boldsymbol{K}_{AA}\right) \tag{8.68}$$

$$\boldsymbol{f}_B \sim \mathscr{N}\left(\boldsymbol{\mu}_B; \boldsymbol{K}_{BB}\right). \tag{8.69}$$

[2]Mathematically, a function can be visualized as a point in an infinite-dimensional space. Without lack of generality, we can restrict this infinite-dimensional space to a Hilbert space.

2. Conditional distributions of a multivariate Gaussian distribution are also Gaussian. The conditional distributions are given by

$$\boldsymbol{f}_A|\boldsymbol{f}_B \sim \mathcal{N}\left(\boldsymbol{\mu}_A + \boldsymbol{K}_{AB}\boldsymbol{K}_{BB}^{-1}\left(\boldsymbol{f}_B - \boldsymbol{\mu}_B\right); \boldsymbol{K}_{AA} - \boldsymbol{K}_{AB}\boldsymbol{K}_{BB}^{-1}\boldsymbol{K}_{BA}\right) \quad (8.70)$$
$$\boldsymbol{f}_B|\boldsymbol{f}_A \sim \mathcal{N}\left(\boldsymbol{\mu}_B + \boldsymbol{K}_{BA}\boldsymbol{K}_{AA}^{-1}\left(\boldsymbol{f}_A - \boldsymbol{\mu}_A\right); \boldsymbol{K}_{BB} - \boldsymbol{K}_{BA}\boldsymbol{K}_{AA}^{-1}\boldsymbol{K}_{AB}\right). \quad (8.71)$$

Using these two properties, we now demonstrate how the Gaussian process can be used for machine learning. Let $\{\boldsymbol{X}; \boldsymbol{Y}\}$ be our training dataset. Using the notation defined in Eq. 8.69, we have $A = \boldsymbol{X}$ and $\boldsymbol{f}_A = \boldsymbol{Y}$. Using the same notation, we define B to be a set of all the possible input settings $\boldsymbol{x}$ that are not considered on $\boldsymbol{X}$. Then by using the marginalization property in (8.69), we obtain

$$\boldsymbol{Y} \sim \mathcal{N}\left(\boldsymbol{\mu}_Y; \boldsymbol{K}_{XX}\right). \quad (8.72)$$

Thus using the marginalization property, we have reduced a Gaussian process defined on a infinite-dimensional random function to a multivariate Gaussian distribution defined on a finite-dimensional random vector. In fact, this property motivates an alternate definition of a Gaussian process as a collection of random variables such that any finite combination has a multivariate Gaussian distribution. Using Eqs. 8.65, mean and covariance matrix of Eq. 8.72 are given by

$$\boldsymbol{\mu}_Y = \boldsymbol{\mu}\left(\boldsymbol{X}\right) \quad (8.73)$$
$$\boldsymbol{K}_{XX} = \mathcal{K}\left(\boldsymbol{X}, \boldsymbol{X}\right). \quad (8.74)$$

Next, we use the second property (8.71) to demonstrate use of the Gaussian process for prediction. Let us assume that we are interested in prediction $\boldsymbol{y}^*$ at a point $\boldsymbol{x}^*$. Using the earlier notation, we have $\boldsymbol{B} = \boldsymbol{x}^*$ and $\boldsymbol{f}_B = \boldsymbol{y}^*$. Thus, using the conditional property of the multivariate Gaussian distribution, we obtain

$$\boldsymbol{y}^*|\boldsymbol{Y} \sim \mathcal{N}\left(\boldsymbol{\mu}_*, \boldsymbol{K}_{*,*}\right), \quad (8.75)$$

where $\boldsymbol{\mu}_*$ and $\boldsymbol{K}_{*,*}$ are mean and variance of the conditional distribution. Using (8.71) and (8.65), the mean is given by

$$\boldsymbol{\mu}_* = \boldsymbol{\mu}(\boldsymbol{x}^*) + \mathcal{K}\left(\boldsymbol{x}^*, \boldsymbol{X}\right)\left[\mathcal{K}\left(\boldsymbol{X}, \boldsymbol{X}\right)\right]^{-1}\left(\boldsymbol{Y} - \boldsymbol{\mu}\left(\boldsymbol{X}\right)\right). \quad (8.76)$$

Similarly, the variance is given by

$$\boldsymbol{K}_{*,*} = \mathcal{K}\left(\boldsymbol{x}^*, \boldsymbol{x}^*\right) - \mathcal{K}\left(\boldsymbol{x}^*, \boldsymbol{X}\right)\left[\mathcal{K}\left(\boldsymbol{X}, \boldsymbol{X}\right)\right]^{-1}\mathcal{K}\left(\boldsymbol{X}, \boldsymbol{x}^*\right). \quad (8.77)$$

Note that the mean given by Eq. 8.76 is our prediction at the point $\boldsymbol{x}^*$. In addition to prediction, this formulation also provides variance given by Eq. 8.77 that quantifies uncertainty in our predictions.

To complete our Gaussian process formulation for machine learning, we need to provide a form for the mean and the covariance function defined in Eq. 8.65. In this chapter, we will restrict our discussion for a specific form of the mean and covariance function. For mean, we use the following linear form:

$$\boldsymbol{\mu}(\boldsymbol{x}) = h(\boldsymbol{x})\boldsymbol{\beta}, \tag{8.78}$$

where $h(\boldsymbol{x})$ is a set of basis functions. For example, we can use $h(\boldsymbol{x}) = \begin{bmatrix} 1 & \boldsymbol{x} & \boldsymbol{x}^2 \ldots \end{bmatrix}$ for a polynomial regression. Let $\boldsymbol{H}$ be a matrix with ith row defined by $h(\boldsymbol{x}_i)$, where $\boldsymbol{x}_i$ is ith training data point. We use the following squared exponential form for the covariance function:

$$\mathcal{K}\left(\boldsymbol{x}_i, \boldsymbol{x}_j\right) = \sigma^2 \exp\left(-\boldsymbol{x}_i \Lambda \boldsymbol{x}_j^T\right), \tag{8.79}$$

where Λ is a diagonal matrix. Diagonal elements of Λ are known as the correlation length parameters. Parameters of the mean, $\boldsymbol{\beta}$, and the covariance function, σ^2 and Λ, are unknown to us. For notational convenience, we define $\boldsymbol{\theta} = \left[\boldsymbol{\beta}, \sigma^2, \Lambda\right]$. We can estimate these parameters using the MLE method.

Without assuming a particular functional form, we have reduced the formulation (8.64) to a traditional machine learning problem in the MLE setting. However, for a Gaussian process, $\boldsymbol{\theta}$ are parameters of the probability distribution and not the input–output mapping as in ANN or SVM. To distinguish from the parameters of the input–output mapping, $\boldsymbol{\theta}$ are known as the hyper-parameters of the Gaussian process. Next, we describe a method for estimation of these hyper-parameters.

For a training dataset $\{\boldsymbol{X}; \boldsymbol{Y}\}$, we use the Bayes theorem to formulate the hyper-parameter estimation problem for the Gaussian process as

$$p(\boldsymbol{\theta}|\boldsymbol{Y}, \boldsymbol{X}) \propto p(\boldsymbol{Y}|\boldsymbol{X}, \boldsymbol{\theta})\, p(\boldsymbol{\theta}). \tag{8.80}$$

We use a completely non-informative prior on $\boldsymbol{\theta}$ and take a negative log of the likelihood function, to convert the Bayesian inference problem in Eq. 8.80 to a MLE estimation problem.

Using Eq. 8.72, the likelihood function for the training dataset is given by

$$\begin{aligned} p(\boldsymbol{Y}|\boldsymbol{X}, \boldsymbol{\theta}) &= \mathcal{N}(\boldsymbol{\mu}_Y; \boldsymbol{K}_{XX}) \\ &\propto |\boldsymbol{K}_{XX}|^{-1/2} \exp\left(-\frac{1}{2}(\boldsymbol{Y} - \boldsymbol{H\beta})^T \boldsymbol{K}_{XX}^{-1} (\boldsymbol{Y} - \boldsymbol{H\beta})\right). \end{aligned} \tag{8.81}$$

The negative log-likelihood is given by

$$\mathcal{L}(\boldsymbol{\theta}) = \frac{1}{2}\log(|\boldsymbol{K}_{XX}|) + \frac{1}{2}(\boldsymbol{Y} - \boldsymbol{H\beta})^T \boldsymbol{K}_{XX}^{-1} (\boldsymbol{Y} - \boldsymbol{H\beta}). \tag{8.82}$$

For a given covariance matrix $\boldsymbol{K}_{XX}$, (8.82) is an objective function of a usual least square estimation problem. Thus, the estimate of $\boldsymbol{\beta}$ is given by

$$\hat{\boldsymbol{\beta}} = \left(\boldsymbol{H}^T \boldsymbol{K}_{XX}^{-1} \boldsymbol{H}\right)^{-1} \boldsymbol{H}^T \boldsymbol{K}_{XX}^{-1} \boldsymbol{Y}. \tag{8.83}$$

Note that $\hat{\boldsymbol{\beta}}$ is a generalized least square estimate of $\boldsymbol{\beta}$. For a fixed $\hat{\boldsymbol{\beta}}$, we need derivatives w.r.t. the covariance function parameters to implement the gradient descent algorithm.

For notational convenience, let us define $\boldsymbol{\theta}_C = \left[\sigma^2; \Lambda\right]$. Thus, we obtain the derivative of $\mathscr{L}(\boldsymbol{\theta})$ w.r.t. $\boldsymbol{\theta}_C$ as

$$\frac{\partial}{\partial \boldsymbol{\theta}_C} \left[\mathscr{L}(\boldsymbol{\theta})\right] = \frac{1}{2} \frac{d}{d\boldsymbol{\theta}_C} \log\left[|\boldsymbol{K}_{XX}|\right] + \frac{1}{2} (\boldsymbol{Y} - \boldsymbol{H}\boldsymbol{\beta})^T \left[\frac{d}{d\boldsymbol{\theta}_C} \boldsymbol{K}_{XX}^{-1}\right] (\boldsymbol{Y} - \boldsymbol{H}\boldsymbol{\beta}). \tag{8.84}$$

The first derivative term is given by

$$\frac{d}{d\boldsymbol{\theta}_C} \log\left[|\boldsymbol{K}_{XX}|\right] = Tr\left[\boldsymbol{K}_{XX}^{-1} \frac{d}{d\boldsymbol{\theta}_C} \boldsymbol{K}_{XX}\right]. \tag{8.85}$$

Similarly, the second derivative term is given by

$$\frac{d}{d\boldsymbol{\theta}_C} \boldsymbol{K}_{XX}^{-1} = -\boldsymbol{K}_{XX}^{-1} \left[\frac{d}{d\boldsymbol{\theta}_C} \boldsymbol{K}_{XX}\right] \boldsymbol{K}_{XX}^{-1}. \tag{8.86}$$

For a squared exponential covariance function defined in (8.79), we can represent the covariance function as

$$\boldsymbol{K}_{XX} = \sigma^2 \boldsymbol{R}_{XX}, \tag{8.87}$$

where $\boldsymbol{R}_{XX}$ is a correlation function with (i, j)th element given by $\exp\left(-\boldsymbol{x}_i \Lambda \boldsymbol{x}_j^T\right)$. For this covariance function, we can analytically obtain the MLE estimate of σ^2 as follows. We first consider a partial derivative w.r.t. σ^2. For the first term given by (8.85), we obtain

$$\begin{aligned} \frac{\partial}{\partial \sigma^2} \log\left[|\boldsymbol{K}_{XX}|\right] &= Tr\left[\frac{1}{\sigma^2} \boldsymbol{R}_{XX}^{-1} \frac{\partial \sigma^2}{\partial \sigma^2} \boldsymbol{R}_{XX}\right] \\ &= \frac{N}{\sigma^2}. \end{aligned} \tag{8.88}$$

Similarly, the second term (8.86) becomes

$$\begin{aligned} \frac{\partial}{d\sigma^2} \boldsymbol{K}_{XX}^{-1} &= -\frac{1}{\sigma^2} \boldsymbol{R}_{XX}^{-1} \left[\frac{\partial \sigma^2}{\partial \sigma^2} \boldsymbol{R}_{XX}\right] \frac{1}{\sigma^2} \boldsymbol{R}_{XX}^{-1} \\ &= -\frac{1}{(\sigma^2)^2} \boldsymbol{R}_{XX}^{-1}. \end{aligned} \tag{8.89}$$

Using (8.88) and (8.89), we obtain

$$\frac{\partial \mathcal{L}(\boldsymbol{\theta})}{\partial \sigma^2} = \frac{1}{2}\frac{N}{\sigma^2} - \frac{1}{2}\frac{1}{(\sigma^2)^2}\left(\boldsymbol{Y} - \boldsymbol{H\beta}\right)^T \boldsymbol{R}_{XX}^{-1}\left(\boldsymbol{Y} - \boldsymbol{H\beta}\right). \tag{8.90}$$

Note that the estimate of σ^2 is defined at the optimum given by

$$\frac{\partial \mathcal{L}(\boldsymbol{\theta})}{\partial \sigma^2} = 0. \tag{8.91}$$

Thus, we obtain

$$\begin{aligned} \frac{N}{\hat{\sigma^2}} &= -\frac{1}{(\hat{\sigma^2})^2}\left(\boldsymbol{Y} - \boldsymbol{H\beta}\right)^T \boldsymbol{R}_{XX}^{-1}\left(\boldsymbol{Y} - \boldsymbol{H\beta}\right) \\ \hat{\sigma^2} &= -\frac{1}{N}\left(\boldsymbol{Y} - \boldsymbol{H}\hat{\boldsymbol{\beta}}\right)^T \boldsymbol{R}_{XX}^{-1}\left(\boldsymbol{Y} - \boldsymbol{H}\hat{\boldsymbol{\beta}}\right). \end{aligned} \tag{8.92}$$

Given the estimates $\hat{\boldsymbol{\beta}}$ and $\hat{\sigma^2}$, we finally consider the partial derivative of $\mathcal{L}(\boldsymbol{\theta})$ w.r.t. Λ. Using Eqs. 8.85 and (8.86) in (8.84) for the squared exponential covariance function (8.87), we obtain

$$\frac{\partial}{\partial \Lambda}\left[\mathcal{L}(\boldsymbol{\theta})\right] = \frac{1}{2}Tr\left[\boldsymbol{R}_{XX}^{-1}\frac{\partial}{\partial \Lambda}\boldsymbol{R}_{XX}\right] - \frac{1}{2\hat{\sigma^2}}\left(\boldsymbol{Y} - \boldsymbol{H}\hat{\boldsymbol{\beta}}\right)^T\left[\boldsymbol{R}_{XX}^{-1}\left[\frac{\partial}{\partial \Lambda}\boldsymbol{R}_{XX}\right]\boldsymbol{R}_{XX}^{-1}\right]\left(\boldsymbol{Y} - \boldsymbol{H}\hat{\boldsymbol{\beta}}\right). \tag{8.93}$$

Having obtained the partial derivative for Λ in (8.93), we can use the stochastic gradient descent algorithm to estimate Λ.

Using the development presented in this section, our Gaussian process can be trained using Algorithm (8).

8.6 Recent Advances: Deep Learning Algorithms

The machine learning algorithms that we have considered till this point are known as shallow networks. This nomenclature is used to signify limited number of hidden layers (often 1 or 2) in the artificial neural network, and similarly a single-layer kernel to define SVM or a covariance function of the Gaussian process. One obvious way to improve these algorithms is to use a deep network a stack of shallow networks, popularly known as deep networks. Resulting algorithms are known as deep learning algorithms. Detailed exposition of these deep learning algorithms is beyond the scope of this book. In this section, we will briefly introduce one particular deep learning algorithm, known as *deep belief network* (DBN). However, steps involved in deriving the DBN can be used to develop other deep learning algorithms also.

Basic building block of the DBN is a probabilistic model known as Restricted Boltzmann Machine (RBM). The RBM is a connected network of stochastic binary

Algorithm 8 Mini-batch Training of Gaussian process

Obtain a training dataset $\{\boldsymbol{X}, \boldsymbol{Y}\}$.

Select a functional form for the covariance function.

Specify the loss function depending on type of the problem (regression or classification).

Specify batch size, number of batches, number of epochs, initial guess for parameters $\boldsymbol{\theta}_0 = \{\Lambda_0\}$, learning rate α, and momentum γ.

for `i=1 TO number_of_epochs` **do**

  for `j=0 TO number_of_batches` **do**

    Randomly select a mini-batch $\{\boldsymbol{X}_j, \boldsymbol{Y}_j\}$ of size `batch_size`.

    Use $\{\boldsymbol{X}_j, \boldsymbol{Y}_j\}$ and Λ_j in Eq. 8.83 to obtain .

    Use $\{\boldsymbol{X}_j, \boldsymbol{Y}_j\}$, $\boldsymbol{\theta}_j = \{\Lambda_j\}$ and $\hat{\boldsymbol{\beta}}$ in Eq. 8.92 to obtain $\hat{\sigma^2}$.

    Use $\{\boldsymbol{X}_j, \boldsymbol{Y}_j\}$, $\hat{\boldsymbol{\beta}}$ and $\hat{\sigma^2}$ in Eq. 8.93 to obtain $\frac{\partial}{\partial \Lambda}\left[\mathscr{L}(\boldsymbol{\theta})\right]$.

    Update the velocity vector as

$$\boldsymbol{v}_{j+1} = \gamma \boldsymbol{v}_j + \alpha \frac{\partial}{\partial \Lambda}\left[\mathscr{L}(\boldsymbol{\theta})\right].$$

    Update the parameters as

$$\Lambda_{j+1} = \Lambda_j - \boldsymbol{v}_{j+1}.$$

  end for

  Set $\Lambda_0 = \Lambda_j$

end for

Set $\hat{\Lambda} = \Lambda_j$.

units {0, 1}. A typical RBM is shown in Fig. 8.3, where $\boldsymbol{v}$ denote visible units and $\boldsymbol{h}$ denote hidden units. Probability distribution of the RBM is defined as

$$p(\boldsymbol{v}, \boldsymbol{h}; \boldsymbol{\theta}) = \frac{\exp\left(-E(\boldsymbol{v}, \boldsymbol{h}; \boldsymbol{\theta})\right)}{Z(\boldsymbol{\theta})}, \tag{8.94}$$

where $Z(\boldsymbol{\theta})$ is a normalizing constant given by

$$Z(\boldsymbol{\theta}) = \sum_{\boldsymbol{v}} \sum_{\boldsymbol{h}} \exp\left(-E(\boldsymbol{v}, \boldsymbol{h}; \boldsymbol{\theta})\right). \tag{8.95}$$

Note that form of the probability distribution in Eq. 8.94 is same as the celebrated Boltzmann distribution in the statistical mechanics. Hence, the resultant network is known as the restricted Boltzmann machine. Following the notation from statistical

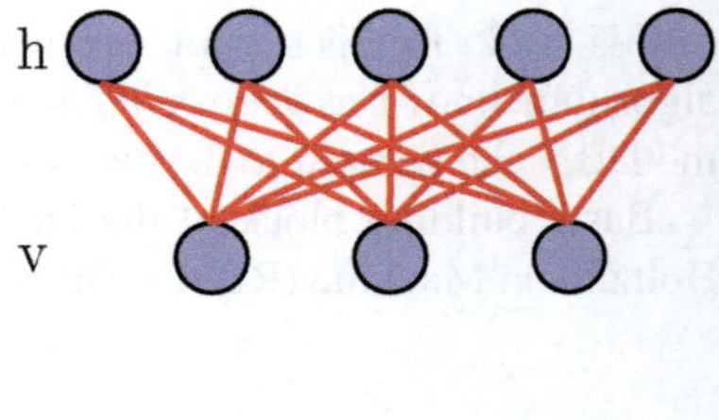

Fig. 8.3 Restricted Boltzmann machine

mechanics, $Z(\boldsymbol{\theta})$ is known as partition function and $E(\boldsymbol{v}, \boldsymbol{h}; \boldsymbol{\theta})$ is known as energy of the system.

For the RBM, energy of the system is defined as

$$E(\boldsymbol{v}, \boldsymbol{h}; \boldsymbol{\theta}) = -\boldsymbol{v}^T \boldsymbol{W} \boldsymbol{h} - \boldsymbol{b}\boldsymbol{v} - \boldsymbol{c}\boldsymbol{h}, \tag{8.96}$$

where $\boldsymbol{W}$ is a matrix of weights, while $\boldsymbol{b}$ and $\boldsymbol{c}$ are bias vectors for the visible and hidden units, respectively. Unknown parameters of the RBM are $\boldsymbol{\theta} = \{\boldsymbol{W}, \boldsymbol{b}, \boldsymbol{c}\}$. We can use the maximum likelihood estimation methodology outlined earlier to estimate the parameters $\boldsymbol{\theta}$. However, there is one key difference between RBM and the machine learning algorithms considered earlier. The RBM is used for the unsupervised learning, that is, we are provided with the input data; however, corresponding labels are not available.

Let $\boldsymbol{V}$ be the dataset available for training. Assuming i.i.d. samples, probability of the dataset is given by

$$p(\{\boldsymbol{V}\}; \boldsymbol{\theta}) = \prod_i \sum_{\boldsymbol{h}} \frac{\exp(-E(\boldsymbol{v}_i, \boldsymbol{h}; \boldsymbol{\theta}))}{Z(\boldsymbol{\theta})}. \tag{8.97}$$

Using Eq. 8.97, the negative log-likelihood is given by

$$\begin{aligned}
\mathscr{L}(\boldsymbol{\theta}) &= -\log\left(\prod_i \sum_{\boldsymbol{h}} \frac{\exp(-E(\boldsymbol{v}_i, \boldsymbol{h}; \boldsymbol{\theta}))}{Z(\boldsymbol{\theta})}\right) \\
&= -\sum_i \left[\log\left(\sum_{\boldsymbol{h}} \exp(-E(\boldsymbol{v}_i, \boldsymbol{h}; \boldsymbol{\theta}))\right) - \log(Z(\boldsymbol{\theta}))\right] \\
&= N \log(Z(\boldsymbol{\theta})) - \sum_i \left[\log\left(\sum_{\boldsymbol{h}} \exp(-E(\boldsymbol{v}_i, \boldsymbol{h}; \boldsymbol{\theta}))\right)\right].
\end{aligned} \tag{8.98}$$

Without loss of generality and some abuse of notation, we divide both sides of Eq. 8.98 by N to obtain

$$\mathscr{L}(\boldsymbol{\theta}) = \log(Z(\boldsymbol{\theta})) - \frac{1}{N} \sum_i \left[\log\left(\sum_{\boldsymbol{h}} \exp(-E(\boldsymbol{v}_i, \boldsymbol{h}; \boldsymbol{\theta}))\right)\right]. \tag{8.99}$$

We can use the mini-batch stochastic gradient descent algorithm to minimize the negative log-likelihood defined in Eq. 8.99; however, first we need to obtain the gradients. Differentiate Eq. 8.99 with respect to $\boldsymbol{\theta}$ to obtain

$$\frac{\partial \mathscr{L}(\boldsymbol{\theta})}{\partial \boldsymbol{\theta}} = \frac{\partial \log(Z(\boldsymbol{\theta}))}{\partial \boldsymbol{\theta}} - \frac{1}{N} \sum_i \left[\frac{\partial \log\left(\sum_{\boldsymbol{h}} \exp(-E(\boldsymbol{v}_i, \boldsymbol{h}; \boldsymbol{\theta}))\right)}{\partial \boldsymbol{\theta}}\right]. \tag{8.100}$$

The second term in Eq. 8.100 is the mean of $\frac{\partial \log\left(\sum_h \exp(-E(\boldsymbol{v}_i, \boldsymbol{h}; \boldsymbol{\theta}))\right)}{\partial \boldsymbol{\theta}}$ obtained using dataset $\boldsymbol{V}$. We denote this mean by

$$\frac{1}{N}\sum_i \left[\frac{\partial \log\left(\sum_h \exp\left(-E(\boldsymbol{v}_i, \boldsymbol{h}; \boldsymbol{\theta})\right)\right)}{\partial \boldsymbol{\theta}}\right] = \left\langle\left[\frac{\partial \log\left(\sum_h \exp\left(-E(\boldsymbol{v}_i, \boldsymbol{h}; \boldsymbol{\theta})\right)\right)}{\partial \boldsymbol{\theta}}\right]\right\rangle_{\boldsymbol{V}}, \quad (8.101)$$

where subscript $\boldsymbol{V}$ highlights the fact that the mean is obtained using the dataset $\boldsymbol{V}$.

The first term in Eq. 8.101 is given by

$$\frac{\partial \log\left(Z(\boldsymbol{\theta})\right)}{\partial \boldsymbol{\theta}} = \frac{1}{Z(\boldsymbol{\theta})}\frac{\partial Z(\boldsymbol{\theta})}{\partial \boldsymbol{\theta}}. \quad (8.102)$$

Using the definition of partition function given by Eq. 8.95 in Eq. 8.102, we obtain

$$\begin{aligned}\frac{\partial \log\left(Z(\boldsymbol{\theta})\right)}{\partial \boldsymbol{\theta}} &= \frac{1}{Z(\boldsymbol{\theta})}\frac{\partial \sum_v \sum_h \exp\left(-E(\boldsymbol{v}, \boldsymbol{h}; \boldsymbol{\theta})\right)}{\partial \boldsymbol{\theta}}\\ &= \sum_v\left[\frac{1}{Z(\boldsymbol{\theta})}\frac{\partial \sum_h \exp\left(-E(\boldsymbol{v}, \boldsymbol{h}; \boldsymbol{\theta})\right)}{\partial \boldsymbol{\theta}}\right]\\ &= \sum_v\left[\frac{\sum_h \exp\left(-E(\boldsymbol{v}, \boldsymbol{h}; \boldsymbol{\theta})\right)}{Z(\boldsymbol{\theta})}\frac{\partial \log\left(\sum_h \exp\left(-E(\boldsymbol{v}, \boldsymbol{h}; \boldsymbol{\theta})\right)\right)}{\partial \boldsymbol{\theta}}\right]\\ &= \sum_v\left[\frac{\partial \log\left(\sum_h \exp\left(-E(\boldsymbol{v}, \boldsymbol{h}; \boldsymbol{\theta})\right)\right)}{\partial \boldsymbol{\theta}}p(\boldsymbol{v}; \boldsymbol{\theta})\right], \end{aligned} \quad (8.103)$$

where we have used definition of the marginal probability distribution to obtain

$$p(\boldsymbol{v}; \boldsymbol{\theta}) = \frac{\sum_h \exp\left(-E(\boldsymbol{v}, \boldsymbol{h}; \boldsymbol{\theta})\right)}{Z(\boldsymbol{\theta})}. \quad (8.104)$$

We recall definition of the expectation operator to note that

$$\sum_v\left[\frac{\partial \log\left(\sum_h \exp\left(-E(\boldsymbol{v}, \boldsymbol{h}; \boldsymbol{\theta})\right)\right)}{\partial \boldsymbol{\theta}}p(\boldsymbol{v}; \boldsymbol{\theta})\right] = \left\langle\left[\frac{\partial \log\left(\sum_h \exp\left(-E(\boldsymbol{v}_i, \boldsymbol{h}; \boldsymbol{\theta})\right)\right)}{\partial \boldsymbol{\theta}}\right]\right\rangle_{p(\boldsymbol{v};\boldsymbol{\theta})}, \quad (8.105)$$

where we have used the subscript $p(\boldsymbol{v}; \boldsymbol{\theta})$ to highlight the fact that the mean is obtained with respect to the true probability distribution $p(\boldsymbol{v}; \boldsymbol{\theta})$. Thus, the gradient is given by

$$\frac{\partial \log\left(Z(\boldsymbol{\theta})\right)}{\partial \boldsymbol{\theta}} = \left\langle\left[\frac{\partial \log\left(\sum_h \exp\left(-E(\boldsymbol{v}_i, \boldsymbol{h}; \boldsymbol{\theta})\right)\right)}{\partial \boldsymbol{\theta}}\right]\right\rangle_{p(\boldsymbol{v};\boldsymbol{\theta})} - \left\langle\left[\frac{\partial \log\left(\sum_h \exp\left(-E(\boldsymbol{v}_i, \boldsymbol{h}; \boldsymbol{\theta})\right)\right)}{\partial \boldsymbol{\theta}}\right]\right\rangle_{\boldsymbol{V}}. \quad (8.106)$$

We can use Eq. 8.106 to evaluate the gradient; however, first we need to resolve following two obstacles:

- Expectation operator in the first term of Eq. 8.106 is defined with respect to $p(\boldsymbol{v}; \boldsymbol{\theta})$. However, $p(\boldsymbol{v}; \boldsymbol{\theta})$ is not known to us.
- We need to evaluate the gradient $\frac{\partial \log\left(\sum_{\boldsymbol{h}} \exp(-E(\boldsymbol{v}_i, \boldsymbol{h}; \boldsymbol{\theta}))\right)}{\partial \boldsymbol{\theta}}$.

The first obstacle is resolved as follows. Evaluation of the expectation with respect to an arbitrary probability distribution $p(\boldsymbol{v}; \boldsymbol{\theta})$ is numerically intractable. However, a sample from the distribution $p(\boldsymbol{v}; \boldsymbol{\theta})$ can be used to approximate the expectation with high accuracy. Markov Chain Monte Carlo (MCMC) method can be used to obtain samples from the distribution $p(\boldsymbol{v}; \boldsymbol{\theta})$. To ensure convergence, the MCMC method requires a large number of transition steps before initiation of sampling. The MCMC can be initialized from the data sample $\{\boldsymbol{V}\}$, and the Gibbs sampling can be used (theoretically infinite number of times) to obtain the sample $\{\boldsymbol{V}\}^{\infty}$ from $p(\boldsymbol{v}; \boldsymbol{\theta})$. Although the MCMC sampling requires a large number of transitions, in practice only few (often a single) step of MCMC can approximate the gradient (8.106) with high accuracy. Using this approximation, the gradient in Eq. 8.106 is given by

$$\frac{\partial \log(\mathscr{L}(\boldsymbol{\theta}))}{\partial \boldsymbol{\theta}} = \left\langle \left[\frac{\partial \log\left(\sum_{\boldsymbol{h}} \exp(-E(\boldsymbol{v}_i, \boldsymbol{h}; \boldsymbol{\theta}))\right)}{\partial \boldsymbol{\theta}} \right] \right\rangle_{\boldsymbol{V}^1} - \left\langle \left[\frac{\partial \log\left(\sum_{\boldsymbol{h}} \exp(-E(\boldsymbol{v}_i, \boldsymbol{h}; \boldsymbol{\theta}))\right)}{\partial \boldsymbol{\theta}} \right] \right\rangle_{\boldsymbol{V}}, \tag{8.107}$$

where the subscript $\boldsymbol{V}^1$ is used to denote the MCMC sample after one transition. Gradient evaluation using Eq. 8.107 is known as the *contrastive-divergence* algorithm.

To resolve the second obstacle, consider the derivative of the logarithmic function as follows:

$$\begin{aligned}
\frac{\partial \log\left(\sum_{\boldsymbol{h}} \exp(-E(\boldsymbol{v}_i, \boldsymbol{h}; \boldsymbol{\theta}))\right)}{\partial \boldsymbol{\theta}} &= \frac{1}{\sum_{\boldsymbol{h}} \exp(-E(\boldsymbol{v}, \boldsymbol{h}; \boldsymbol{\theta}))} \sum_{\boldsymbol{h}} \frac{\partial \exp(-E(\boldsymbol{v}, \boldsymbol{h}; \boldsymbol{\theta}))}{\partial \boldsymbol{\theta}} \\
&= \sum_{\boldsymbol{h}} \frac{\exp(-E(\boldsymbol{v}, \boldsymbol{h}; \boldsymbol{\theta}))}{\sum_{\boldsymbol{h}} \exp(-E(\boldsymbol{v}, \boldsymbol{h}; \boldsymbol{\theta}))} \frac{\partial(-E(\boldsymbol{v}, \boldsymbol{h}; \boldsymbol{\theta}))}{\partial \boldsymbol{\theta}} \\
&= -\sum_{\boldsymbol{h}} \frac{\exp(-E(\boldsymbol{v}, \boldsymbol{h}; \boldsymbol{\theta}))}{\sum_{\boldsymbol{h}} \exp(-E(\boldsymbol{v}, \boldsymbol{h}; \boldsymbol{\theta}))} \frac{\partial(E(\boldsymbol{v}, \boldsymbol{h}; \boldsymbol{\theta}))}{\partial \boldsymbol{\theta}} \\
&= -\sum_{\boldsymbol{h}} p(\boldsymbol{h}|\boldsymbol{v}; \boldsymbol{\theta}) \frac{\partial(E(\boldsymbol{v}, \boldsymbol{h}; \boldsymbol{\theta}))}{\partial \boldsymbol{\theta}}.
\end{aligned} \tag{8.108}$$

Collating the development till this point, we obtain the required gradient as

$$\frac{\partial \mathscr{L}(\boldsymbol{\theta})}{\partial \boldsymbol{\theta}} = \left\langle \sum_{\boldsymbol{h}} p(\boldsymbol{h}|\boldsymbol{v}; \boldsymbol{\theta}) \frac{\partial(E(\boldsymbol{v}, \boldsymbol{h}; \boldsymbol{\theta}))}{\partial \boldsymbol{\theta}} \right\rangle_{\boldsymbol{V}} - \left\langle \sum_{\boldsymbol{h}} p(\boldsymbol{h}|\boldsymbol{v}; \boldsymbol{\theta}) \frac{\partial(E(\boldsymbol{v}, \boldsymbol{h}; \boldsymbol{\theta}))}{\partial \boldsymbol{\theta}} \right\rangle_{\boldsymbol{V}^1}. \tag{8.109}$$

Note that Eq. 8.109 only involves evaluation of gradient of the energy function, which is a linear function of parameters. For energy function of the RBM defined in Eq. 8.96, the gradients are given by

$$\begin{aligned}\frac{\partial\left(E(\boldsymbol{v},\boldsymbol{h};\boldsymbol{\theta})\right)}{\partial\boldsymbol{W}} &= -\boldsymbol{v}^T\boldsymbol{h}\\ \frac{\partial\left(E(\boldsymbol{v},\boldsymbol{h};\boldsymbol{\theta})\right)}{\partial\boldsymbol{b}} &= -\boldsymbol{v}\\ \frac{\partial\left(E(\boldsymbol{v},\boldsymbol{h};\boldsymbol{\theta})\right)}{\partial\boldsymbol{c}} &= -\boldsymbol{h}.\end{aligned} \tag{8.110}$$

Using Eq. 8.110 in Eq. 8.109 and noting that $\boldsymbol{v}$ and $\boldsymbol{h}$ are binary random variables taking values in $\{0, 1\}$, we obtain

$$\begin{aligned}\frac{\partial\mathscr{L}(\boldsymbol{\theta})}{\partial\boldsymbol{W}} &= \langle p(\boldsymbol{h}=1|\boldsymbol{v};\boldsymbol{\theta})\boldsymbol{v}\rangle_{\boldsymbol{V}} - \langle p(\boldsymbol{h}=1|\boldsymbol{v};\boldsymbol{\theta})\boldsymbol{v}\rangle_{\boldsymbol{V}^1}\\ \frac{\partial\mathscr{L}(\boldsymbol{\theta})}{\partial\boldsymbol{b}} &= \langle\boldsymbol{v}\rangle_{\boldsymbol{V}} - \langle\boldsymbol{v}\rangle_{\boldsymbol{V}^1}\\ \frac{\partial\mathscr{L}(\boldsymbol{\theta})}{\partial\boldsymbol{c}} &= \langle p(\boldsymbol{h}=1|\boldsymbol{v};\boldsymbol{\theta})\rangle_{\boldsymbol{V}} - \langle p(\boldsymbol{h}=1|\boldsymbol{v};\boldsymbol{\theta})\rangle_{\boldsymbol{V}^1}.\end{aligned} \tag{8.111}$$

We can train our RBM using Algorithm 9.

Algorithm 9 Mini-batch Training of Restricted Boltzmann Machine

Obtain a training dataset $\{\boldsymbol{V}\}$.
Specify nodes in the hidden layer.
Specify batch size, number of batches, number of epochs, initial guess for parameters $\boldsymbol{\theta}_0$, learning rate α, and momentum γ.
for `i=1 TO number_of_epochs` **do**
  for `j=0 TO number_of_batches` **do**
    Randomly select a mini-batch $\{\boldsymbol{V}_j\}$ of size `batch_size`.
    Sample $\boldsymbol{H} \in p\left(\boldsymbol{h}|\boldsymbol{V}_j, \boldsymbol{\theta}_j\right)$.
    Sample $\boldsymbol{V}^1 \in p\left(\boldsymbol{v}|\boldsymbol{H}, \boldsymbol{\theta}_j\right)$.
    Obtain the gradients using Eq. 8.111
    Update the velocity vector as

$$\boldsymbol{v}_{j+1} = \gamma\boldsymbol{v}_j + \alpha\nabla_{\boldsymbol{\theta}}\mathscr{L}(\boldsymbol{\theta}).$$

    Update the parameters as

$$\boldsymbol{\theta}_{j+1} = \boldsymbol{\theta}_j - \boldsymbol{v}_{j+1}.$$

  end for
  Set $\boldsymbol{\theta}_0 = \boldsymbol{\theta}_j$
end for
Set $\hat{\boldsymbol{\theta}} = \boldsymbol{\theta}_j$.

As described earlier, the RBM is a shallow network. We can create a deep network by stacking the RBMs together. In practice, many RBMs are connected with each other to form a deep network. This deep network is known as a *deep belief network*. For demonstration, let us consider a three-layer DBN with visible layer $\boldsymbol{v}$ and two hidden layers $\boldsymbol{h}_1$ and $\boldsymbol{h}_2$. Following observation motivates the construction of the DBN:

$$p(\boldsymbol{v}, \boldsymbol{h}_1, \boldsymbol{h}_2; \boldsymbol{\theta}) = p(\boldsymbol{v}|\boldsymbol{h}_1; \boldsymbol{\theta}_1)p(\boldsymbol{h}_1, \boldsymbol{h}_2; \boldsymbol{\theta}_2), \tag{8.112}$$

where $\boldsymbol{\theta}_1$ and $\boldsymbol{\theta}_2$ are the parameters of the first and the second RBM, respectively, and $\boldsymbol{\theta} = \{\boldsymbol{\theta}_1, \boldsymbol{\theta}_2\}$. Equation 8.112 shows that the deep network can be constructed by concatenating the RBM ($p(\boldsymbol{h}_1, \boldsymbol{h}_2; \boldsymbol{\theta}_2)$, top two layers), with the bottom conditional distribution ($p(\boldsymbol{v}|\boldsymbol{h}_1; \boldsymbol{\theta}_1)$, also known as the belief network). For a special case of $\boldsymbol{\theta}_1 = \boldsymbol{\theta}_2$, it can show that the bottom two layers also form a RBM.

Construction of the DBN from independent RBMs is exploited for greedy layerwise training of the DBN. Let the bottom RBM is trained using Algorithm (9) to estimate the parameters $\boldsymbol{\theta}_1$. When $\boldsymbol{\theta}_2 = \boldsymbol{\theta}_1$, the DBN is at least as good as the bottom RBM. By adding one more layer $\boldsymbol{h}_2$ and estimating the parameters $\boldsymbol{\theta}_2$, prediction accuracy of the network can be improved.

To explore further, consider

$$\begin{aligned} p(\boldsymbol{v}; \boldsymbol{\theta}) &= \sum_{\boldsymbol{h}_1} \sum_{\boldsymbol{h}_2} p(\boldsymbol{v}, \boldsymbol{h}_1, \boldsymbol{h}_2; \boldsymbol{\theta}) \\ &= \sum_{\boldsymbol{h}_1} \sum_{\boldsymbol{h}_2} p(\boldsymbol{v}|\boldsymbol{h}_1; \boldsymbol{\theta}_1)p(\boldsymbol{h}_1, \boldsymbol{h}_2; \boldsymbol{\theta}_2) \\ &= \sum_{\boldsymbol{h}_1} p(\boldsymbol{v}|\boldsymbol{h}_1; \boldsymbol{\theta}_1) \sum_{\boldsymbol{h}_2} p(\boldsymbol{h}_1, \boldsymbol{h}_2; \boldsymbol{\theta}_2). \end{aligned} \tag{8.113}$$

Note that for a case when $\boldsymbol{\theta}_1 = \boldsymbol{\theta}_2$,

$$\begin{aligned} p(\boldsymbol{v}|\boldsymbol{h}_1; \boldsymbol{\theta}_1) \sum_{\boldsymbol{h}_2} p(\boldsymbol{h}_1, \boldsymbol{h}_2; \boldsymbol{\theta}_2) &= p(\boldsymbol{v}|\boldsymbol{h}_1; \boldsymbol{\theta}_1) \sum_{\boldsymbol{h}_2} p(\boldsymbol{h}_1, \boldsymbol{h}_2; \boldsymbol{\theta}_1) \\ &= p(\boldsymbol{v}|\boldsymbol{h}_1; \boldsymbol{\theta}_1)p(\boldsymbol{h}_1; \boldsymbol{\theta}_1) \\ &= p(\boldsymbol{v}, \boldsymbol{h}_1; \boldsymbol{\theta}_1). \end{aligned} \tag{8.114}$$

Parameters $\boldsymbol{\theta}_1$ are obtained by maximizing the log-likelihood in Eq. 8.114 using Algorithm (9).

Let $\boldsymbol{\theta}_1$ be known from the previous step; reconsider Eq. 8.113:

$$
\begin{aligned}
p(\boldsymbol{v};\boldsymbol{\theta}) &= \sum_{\boldsymbol{h}_1} p(\boldsymbol{v}|\boldsymbol{h}_1;\boldsymbol{\theta}_1) \sum_{\boldsymbol{h}_2} p(\boldsymbol{h}_1,\boldsymbol{h}_2;\boldsymbol{\theta}_2) \\
&= \sum_{\boldsymbol{h}_1} p(\boldsymbol{h}_1|\boldsymbol{v};\boldsymbol{\theta}_1) \frac{p(\boldsymbol{v}|\boldsymbol{h}_1;\boldsymbol{\theta}_1)}{p(\boldsymbol{h}_1|\boldsymbol{v};\boldsymbol{\theta}_1)} \sum_{\boldsymbol{h}_2} p(\boldsymbol{h}_1,\boldsymbol{h}_2;\boldsymbol{\theta}_2) \\
&= \sum_{\boldsymbol{h}_1} p(\boldsymbol{h}_1|\boldsymbol{v};\boldsymbol{\theta}_1) \frac{p(\boldsymbol{v}|\boldsymbol{h}_1;\boldsymbol{\theta}_1) \sum_{\boldsymbol{h}_2} p(\boldsymbol{h}_1,\boldsymbol{h}_2;\boldsymbol{\theta}_2)}{p(\boldsymbol{h}_1|\boldsymbol{v};\boldsymbol{\theta}_1)} \\
&= E_{p(\boldsymbol{h}_1|\boldsymbol{v};\boldsymbol{\theta}_1)} \left(\frac{p(\boldsymbol{v}|\boldsymbol{h}_1;\boldsymbol{\theta}_1) \sum_{\boldsymbol{h}_2} p(\boldsymbol{h}_1,\boldsymbol{h}_2;\boldsymbol{\theta}_2)}{p(\boldsymbol{h}_1|\boldsymbol{v};\boldsymbol{\theta}_1)} \right). \qquad (8.115)
\end{aligned}
$$

Note that in Eq. 8.115, $p(\boldsymbol{h}_1|\boldsymbol{v};\boldsymbol{\theta}_1)$ can be considered as an equivalent of Bayesian prior of $\boldsymbol{h}_1$, which is updated by maximizing the likelihood of $\sum_{\boldsymbol{h}_2} p(\boldsymbol{h}_1,\boldsymbol{h}_2;\boldsymbol{\theta}_2)$. The log-likelihood of Eq. 8.115 is given by

$$
\log\left(p(\boldsymbol{v};\boldsymbol{\theta})\right) = \log\left(E_{p(\boldsymbol{h}_1|\boldsymbol{v};\boldsymbol{\theta}_1)} \left(\frac{p(\boldsymbol{v}|\boldsymbol{h}_1;\boldsymbol{\theta}_1) \sum_{\boldsymbol{h}_2} p(\boldsymbol{h}_1,\boldsymbol{h}_2;\boldsymbol{\theta}_2)}{p(\boldsymbol{h}_1|\boldsymbol{v};\boldsymbol{\theta}_1)} \right)\right). \qquad (8.116)
$$

However, maximization of the log-likelihood in Eq. 8.116 is computationally intractable. Instead in the greedy training algorithm, a lower bound on the log-likelihood is maximized. This lower bound is obtained using the Jensen's inequality.

For a concave function $\phi(\cdot)$, the Jensen's inequality gives

$$
E(\phi(\cdot)) \leq \phi(E(\cdot)). \qquad (8.117)
$$

Since log is a concave function, application of the Jensen's inequality in Eq. 8.116 provides

$$
\begin{aligned}
\log\left(p(\boldsymbol{v};\boldsymbol{\theta})\right) &= \log\left(E_{p(\boldsymbol{h}_1|\boldsymbol{v};\boldsymbol{\theta}_1)} \left(\frac{p(\boldsymbol{v}|\boldsymbol{h}_1;\boldsymbol{\theta}_1) \sum_{\boldsymbol{h}_2} p(\boldsymbol{h}_1,\boldsymbol{h}_2;\boldsymbol{\theta}_2)}{p(\boldsymbol{h}_1|\boldsymbol{v};\boldsymbol{\theta}_1)} \right)\right) \\
&\geq E_{p(\boldsymbol{h}_1|\boldsymbol{v};\boldsymbol{\theta}_1)} \left(\log\left(\frac{p(\boldsymbol{v}|\boldsymbol{h}_1;\boldsymbol{\theta}_1) \sum_{\boldsymbol{h}_2} p(\boldsymbol{h}_1,\boldsymbol{h}_2;\boldsymbol{\theta}_2)}{p(\boldsymbol{h}_1|\boldsymbol{v};\boldsymbol{\theta}_1)} \right)\right) \\
&= \sum_{\boldsymbol{h}_1} p(\boldsymbol{h}_1|\boldsymbol{v};\boldsymbol{\theta}_1) \log\left(\frac{p(\boldsymbol{v}|\boldsymbol{h}_1;\boldsymbol{\theta}_1) \sum_{\boldsymbol{h}_2} p(\boldsymbol{h}_1,\boldsymbol{h}_2;\boldsymbol{\theta}_2)}{p(\boldsymbol{h}_1|\boldsymbol{v};\boldsymbol{\theta}_1)} \right) \\
&= \sum_{\boldsymbol{h}_1} p(\boldsymbol{h}_1|\boldsymbol{v};\boldsymbol{\theta}_1) \log\left(p(\boldsymbol{v}|\boldsymbol{h}_1;\boldsymbol{\theta}_1) \sum_{\boldsymbol{h}_2} p(\boldsymbol{h}_1,\boldsymbol{h}_2;\boldsymbol{\theta}_2) \right) - \\
&\quad \sum_{\boldsymbol{h}_1} p(\boldsymbol{h}_1|\boldsymbol{v};\boldsymbol{\theta}_1) \log\left(p(\boldsymbol{h}_1|\boldsymbol{v};\boldsymbol{\theta}_1)\right). \qquad (8.118)
\end{aligned}
$$

Use the definition of the Shannon's entropy

$$
\mathscr{H}\left(p(\boldsymbol{h}_1|\boldsymbol{v};\boldsymbol{\theta}_1)\right) = -\sum_{\boldsymbol{h}_1} p(\boldsymbol{h}_1|\boldsymbol{v};\boldsymbol{\theta}_1) \log\left(p(\boldsymbol{h}_1|\boldsymbol{v};\boldsymbol{\theta}_1)\right), \qquad (8.119)
$$

in Eq. 8.118 to obtain

$$\begin{aligned}\log\left(p(\boldsymbol{v};\boldsymbol{\theta})\right) &\geq \sum_{\boldsymbol{h}_1} p(\boldsymbol{h}_1|\boldsymbol{v};\boldsymbol{\theta}_1)\log\left(p(\boldsymbol{v}|\boldsymbol{h}_1;\boldsymbol{\theta}_1)\sum_{\boldsymbol{h}_2} p(\boldsymbol{h}_1,\boldsymbol{h}_2;\boldsymbol{\theta}_2)\right) - \mathscr{H}(p(\boldsymbol{h}_1|\boldsymbol{v};\boldsymbol{\theta}_1))\\ &= \sum_{\boldsymbol{h}_1} p(\boldsymbol{h}_1|\boldsymbol{v};\boldsymbol{\theta}_1)\log(p(\boldsymbol{v}|\boldsymbol{h}_1;\boldsymbol{\theta}_1))\\ &\quad + \sum_{\boldsymbol{h}_1} p(\boldsymbol{h}_1|\boldsymbol{v};\boldsymbol{\theta}_1)\log\left(\sum_{\boldsymbol{h}_2} p(\boldsymbol{h}_1,\boldsymbol{h}_2;\boldsymbol{\theta}_2)\right)\\ &\quad - \mathscr{H}(p(\boldsymbol{h}_1|\boldsymbol{v};\boldsymbol{\theta}_1)).\end{aligned} \tag{8.120}$$

Thus for a fixed $\boldsymbol{\theta}_1$, maximization of the lower bound on $\log\left(p(\boldsymbol{v};\boldsymbol{\theta})\right)$ is equivalent to the solution of the following optimization problem:

$$\begin{aligned}\hat{\boldsymbol{\theta}_2} &= \arg\max_{\boldsymbol{\theta}_2}\sum_{\boldsymbol{h}_1} p(\boldsymbol{h}_1|\boldsymbol{v};\boldsymbol{\theta}_1)\log\left(\sum_{\boldsymbol{h}_2} p(\boldsymbol{h}_1,\boldsymbol{h}_2;\boldsymbol{\theta}_2)\right)\\ &= \arg\max_{\boldsymbol{\theta}_2}\sum_{\boldsymbol{h}_1} p(\boldsymbol{h}_1|\boldsymbol{v};\boldsymbol{\theta}_1)\log\left(p(\boldsymbol{h}_1;\boldsymbol{\theta}_2)\right)\\ &= \sum_{\boldsymbol{h}_1} p(\boldsymbol{h}_1|\boldsymbol{v};\boldsymbol{\theta}_1)\left(\arg\max_{\boldsymbol{\theta}_2}\left(\log\left(p(\boldsymbol{h}_1;\boldsymbol{\theta}_2)\right)\right)\right)\\ &= E_{p(\boldsymbol{h}_1|\boldsymbol{v};\boldsymbol{\theta}_1)}\left(\arg\max_{\boldsymbol{\theta}_2}\quad\log\left(p(\boldsymbol{h}_1;\boldsymbol{\theta}_2)\right)\right).\end{aligned} \tag{8.121}$$

Note that solution of

$$\arg\max_{\boldsymbol{\theta}_2}\quad\log\left(p(\boldsymbol{h}_1;\boldsymbol{\theta}_2)\right), \tag{8.122}$$

is numerically obtained using Algorithm (9). The expectation in Eq. 8.121 is obtained by generating a dataset $\boldsymbol{H}^1 \sim p(\boldsymbol{h}_1|\boldsymbol{V}^0;\boldsymbol{\theta}_1)$, and training the RBM $p(\boldsymbol{h}_1,\boldsymbol{h}_2;\boldsymbol{\theta}_2)$ using this dataset. We conclude this section with the method for training the DBN as summarized in Algorithm (10).

Algorithm 10 Training the Deep Belief Network

1: Obtain a training dataset $\{\boldsymbol{V}\}$.
2: Specify number of hidden layers and number of nodes in each hidden layer.
3: Train the first RBM using the training dataset $\{\boldsymbol{V}\}$.
4: Generate $\boldsymbol{H}^1 \in p\,(\boldsymbol{h}|\boldsymbol{V};\boldsymbol{\theta}_1)$.
5: **for** `i=2 TO number_of_hidden_layers` **do**
6: Train ith RBM using the data $\boldsymbol{H}^{i-1}$.
7: Generate the data $\boldsymbol{H}^i \in p\,(\boldsymbol{h}_i|\boldsymbol{h}_{i-1};\boldsymbol{\theta}_1)$.
8: **end for**

Chapter 9
Key Applications of State and Health Estimation

In the previous two chapters, we have developed theoretical foundation of data-driven methods for lithium-ion batteries. In this chapter, we present test cases to demonstrate applicability and capabilities of these methods for lithium-ion battery state estimation. First, we explore the recursive Bayesian framework for state of charge estimation. In this chapter, we compare the unscented Kalman filter and particle filter for state of charge estimation. Functionality of these algorithms is demonstrated for a commercial NCA/C cell state estimation at different operating conditions including constant current discharge at room and low temperatures, hybrid power pulse characterization (HPPC), and urban driving schedule (UDDS) protocols. In addition to accurate voltage prediction, the electrochemical nature of ROM enables drawing of physical insights into the cell behavior. Advantages of using electrode concentrations over conventional Coulomb counting for accessible capacity estimation are discussed. In addition to the mean state estimation, the framework also provides estimation of the associated confidence bounds that are used to establish predictive capability of the proposed framework. Next, we demonstrate applicability of the machine learning algorithms for lithium-ion battery state of health estimation. For this, we present a novel method that utilizes both the classification and regression flavors of the machine learning algorithms. For demonstration purpose, we consider SVM for classification and regression; however, other approaches can be similarly used. For this demonstration, we have used a publicly available battery life testing dataset for training the SVM/R algorithm and subsequently tested our approach on a different subset of the dataset.[1]

---

[1]Figures and discussions reproduced with permissions from Elsevier.

K. S. Hariharan et al., *Mathematical Modeling of Lithium Batteries*,
Green Energy and Technology, https://doi.org/10.1007/978-3-319-03527-7_9

9.1 Li-Ion Battery State Estimation Using the Recursive Bayesian Filter

In the last part of the book, we have developed a reduced-order model of the Li-ion battery by using volume averaging of the P2D-ECT model and associated profile approximations. This reduced-order model is computationally more efficient than the P2D-ECT model, and thus is amenable for implementation of onboard BMS. The BMS is also equipped with a sensor system that continuously monitors cell voltage, current, and temperature. We use the recursive Bayesian framework developed earlier to assimilate sensor data with the reduced-order model for accurate state estimation.

First step in the implementation of the RBF is to develop a state-space representation of the ROM model. Using the nomenclature in Eq. 7.9, the state is given by[2]

$$\mathbf{x} = \left[c_{1n}, c_{1p}, c_{1rn}, c_{1rp}, q_{2in}, q_{2ip}\right]. \tag{9.1}$$

Dynamics of these six state variables is given by the governing equations of the ROM model. For completeness, the governing equations of the ROM are summarized in Table 9.1.

Note that the governing equations for the interfacial flux are coupled. We will decouple these governing equations to obtain the state-space form. Governing equations for the interfacial flux can be written as

$$\begin{aligned} A_1\frac{dq_{2in}}{dt} + A_2\frac{dq_{2ip}}{dt} &= -q_{2in} + A_3\left\langle j_n\right\rangle \\ B_1\frac{dq_{2in}}{dt} + B_2\frac{dq_{2ip}}{dt} &= q_{2ip} + B_3\left\langle j_p\right\rangle. \end{aligned} \tag{9.2}$$

Multiply the first equation in Eq. 9.2 by B_2 and the second equation by A_2 to obtain

$$\begin{aligned} A_1B_2\frac{dq_{2in}}{dt} + A_2B_2\frac{dq_{2ip}}{dt} &= -B_2q_{2in} + A_3B_2\left\langle j_n\right\rangle \\ A_2B_1\frac{dq_{2in}}{dt} + A_2B_2\frac{dq_{2ip}}{dt} &= A_2q_{2ip} + A_2B_3\left\langle j_p\right\rangle. \end{aligned} \tag{9.3}$$

If we subtract the second equation of Eq. 9.3 from the first, we obtain

$$\frac{dq_{2in}}{dt} = -\frac{B_2}{D_1}q_{2in} - \frac{A_2}{D_1}q_{2ip} + \frac{A_3B_2}{D_1}\left\langle j_n\right\rangle - \frac{A_2B_3}{D_1}\left\langle j_p\right\rangle, \tag{9.4}$$

where

$$D_1 = A_1B_2 - A_2B_1. \tag{9.5}$$

[2]Please refer Chap. 5 for nomenclature.

Table 9.1 Table gives a compilation of governing equations for the reduced-order model

Negative electrode	Positive electrode
$\frac{d\langle\bar{c}_{1n}\rangle}{dt} = -\frac{3\langle j_n\rangle}{r_n}$	$\frac{d\langle\bar{c}_{1p}\rangle}{dt} = -\frac{3\langle j_p\rangle}{r_p}$
$\frac{d\langle\bar{c}_{1rn}\rangle}{dt} = -\frac{30D_{1n}\langle\bar{c}_{1rn}\rangle}{R_n^2} - \frac{45\langle j_n\rangle}{2R_n^2}$	$\frac{d\langle\bar{c}_{1rp}\rangle}{dt} = -\frac{30D_{1p}\langle\bar{c}_{1rp}\rangle}{R_p^2} - \frac{45\langle j_p\rangle}{2R_p^2}$
Interfacial Flux	
$\left(l_n\varepsilon_{2n}\alpha_{in} + \frac{l_s l_n \varepsilon_{2n}}{2D_{2s}} + \frac{l_n^2\varepsilon_{2n}}{3D_{2n}}\right)\frac{dq_{2in}}{dt} + \left(l_n\varepsilon_{2n}\alpha_{ip} + \frac{l_n l_s \varepsilon_{2n}}{2D_{2s}}\right)\frac{dq_{2ip}}{dt} = -q_{2in} + (1-t_+)\,a_n l_n\,\langle j_n\rangle$	
$\left(l_p\varepsilon_{2p}\alpha_{in}\right)\frac{dq_{2in}}{dt} + \left(l_p\varepsilon_{2p}\alpha_{ip} - \frac{l_p^2\varepsilon_{2p}}{3D_{2p}}\right)\frac{dq_{2ip}}{dt} = q_{2ip} + (1-t_+)\,a_p l_p\,\langle j_p\rangle$	

Using the similar procedure, we obtain

$$\frac{dq_{2ip}}{dt} = \frac{B_1}{D_1}q_{2in} + \frac{A_1}{D_1}q_{2ip} - \frac{A_3B_1}{D_1}\langle j_n\rangle + \frac{A_1B_3}{D_1}\langle j_p\rangle. \tag{9.6}$$

Thus, the ROM can be represented in a differential form at the time instance k as

$$\frac{d\boldsymbol{x}_k}{dt} = \boldsymbol{F}\boldsymbol{x}_k + \boldsymbol{B}\boldsymbol{u}_k. \tag{9.7}$$

By using an appropriate numerical integration technique, state-space form for the model is recovered. For example, we can use a simple explicit Euler's method to obtain

$$\boldsymbol{x}_k = \frac{d\boldsymbol{x}_{k-1}}{dt}\Delta T, \tag{9.8}$$

where ΔT is the time step.

For demonstration purpose, we compare the UKF and PF algorithms for numerical implementation of the RBF. Using UKF and PF, the RBF can be implemented using Algorithm 11.

Algorithm 11 Physics-based SOC estimation

1: Initialize the framework using $\mathbf{x}_0$, specify $p(\mathbf{x}_0)$ using sigma vectors for UKF or particles for PF
2: Propagate the uncertainty forward by simulating the model for each sigma vector/particle
3: At an arbitrary time step k, use UKF/PF data assimilation step to estimate $p(\mathbf{x}_k|\mathbf{y}_{1:k})$, where $\mathbf{y}_{1:k}$ is monitored cell voltage.
4: Estimate $p(\mathbf{z}_k|\mathbf{x}_k)$, where $\mathbf{z}_k$ is an intermediate cell variable.
5: Estimate the confidence bound $\mathbf{I}_z^\delta = \{\mathbf{z}_k;\quad p(\mathbf{z}_k) \geq \delta\ \&\ p(\mathbf{z}_k) < (1-\delta)\}$
6: Use uncertainty propagation to estimate $p(\mathbf{x}_{k+1:T}|\mathbf{x}_{0:k}, \mathbf{y}_{1:k})$, $p(\mathbf{z}_{k+1:T}|\mathbf{x}_{k+1:T})$ and the associated confidence bounds.

Effectiveness of the proposed framework is demonstrated for state estimation of the NCA/C cell. The NCA/C cell is a second generation, 18650 size commercial cell with a maximum capacity of 3.3Ah. The cell was tested at room temperatures ($T = 298K$) and low temperatures ($T = 273K$, $263K$ and $253K$) with different

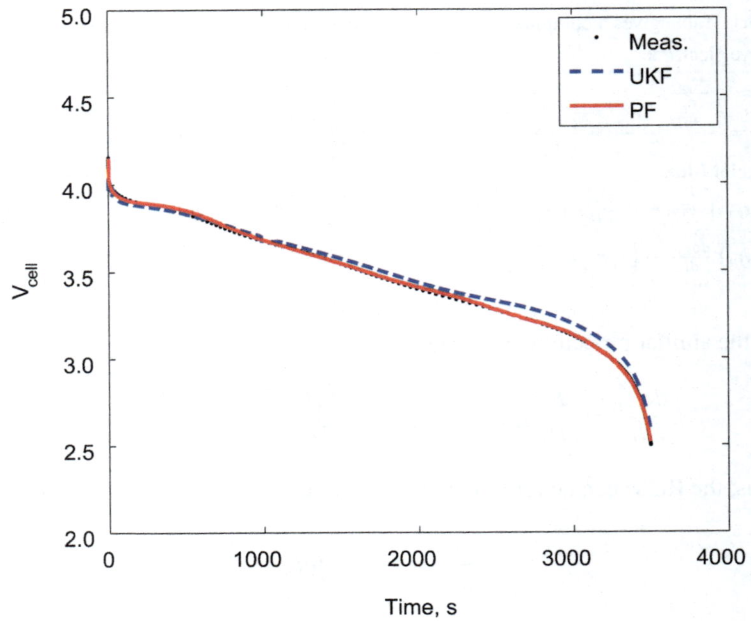

Fig. 9.1 Comparison of voltage estimate using UKF and PF

constant current discharge protocols (C/1, C/2, C/10) in a controlled instrumented laboratory setup. The cell was also tested for hybrid pulse power protocol (HPPC) and urban dynamometer driving schedule (UDDS) protocols at the room temperature. The end of discharge was specified at the cell voltage of 2.4V. Load current and cell voltage were monitored at nonuniform frequency for all the tests. For the test cases presented in this chapter, monitored load current is used as a control input and the measured cell voltage is assimilated for state estimation.

For all the test cases presented in this chapter, the cell is initialized from an equilibrium condition, while the initial electrode surface lithium concentration is assumed uncertain. The initial states are assumed independent and the uncertainty is specified using a Gaussian distribution with mean $\mathbf{x}_0$ and the standard deviation $\sigma_0 = 0.2\mathbf{x}_0$. Uncertainty in the cell voltage measurement is quantified using a zero-mean Gaussian distribution with a standard deviation $\sigma_V = 0.05V$ for all the test cases.

In the first set of simulations, functionality of the proposed framework is investigated for a cell tested at the room temperature. The first test case presents voltage estimation results for a constant current discharge at $C/1$ rate. For this test case, the ROM is simulated using a forward in time Euler solver with time step $\Delta T = 1s$, while the nonuniformly sampled cell voltage data is used for the data assimilation. Cell voltage estimation results are presented in Fig. 9.1. The figure compares cell voltage estimates using UKF (blue line) and PF (red line) against the measurements (black dots). Posterior mean is used as an estimate of the truth for UKF and PF.

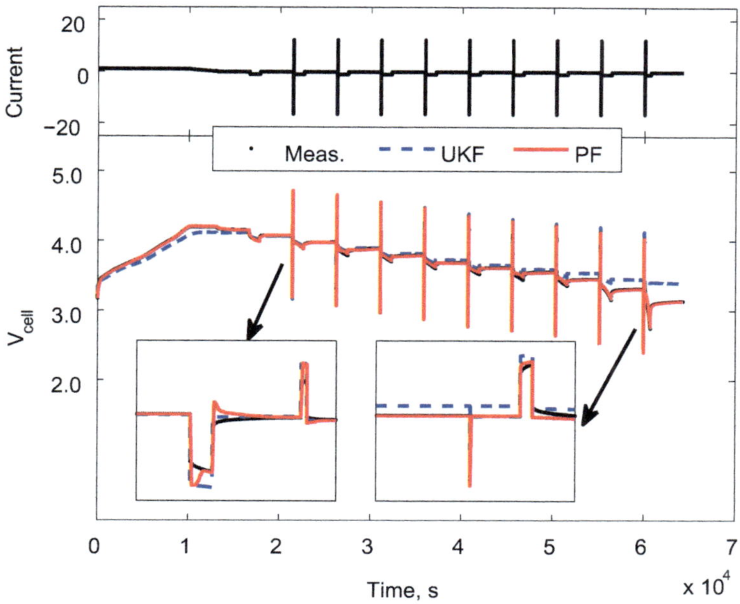

Fig. 9.2 Comparison UKF and PF for HPPC protocol

The UKF cell voltage estimate is close to the measurements; however, the match is poor near the end of discharge (EOD). The PF estimate matches closely with the measurement throughout the discharge cycle. The maximum deviation of the UKF estimate from measurement is about $0.07V$ while for the PF estimate it is about $0.01V$.

Estimation results for the HPPC are presented in Fig. 9.2. Frame (a) of the figure shows estimated cell voltage comparison and frame (b) shows the root-mean-squared error (RMSE) for UKF and PF estimates. The top panel of the frame (a) shows the pulse protocol used for this test. The cell was charged at C/3 rate for 3 hours, followed by a rest of 100 minutes. Subsequently, the cell was subjected to a pulse power protocol of 4 C pulse followed by a 20-minute C/3 discharge and 60 minutes of rest. Duration of HPPC is 18 hours with the data procured at every second. Bottom panel of the frame (a) shows the estimated cell voltage comparison with the measurements. The cell voltage estimated using UKF and PF matches closely with the measurements at high state of charge, while near the low state of charge, the PF estimate matches closer with the measurements as compared to the UKF. The cell voltage comparison at the high and low states of charge is shown in the inset.

The cell was also tested for the UDDS, which is an Environmental Protection Agency mandated test protocol for the urban driving conditions. Figure 9.3 presents state estimation results for the UDDS protocol. Estimated cell voltage comparison is shown in frame (a) and the RMSE comparison for UKF and PF is shown in frame (b). Top panel of the frame (a) shows UDDS profile used for the present test

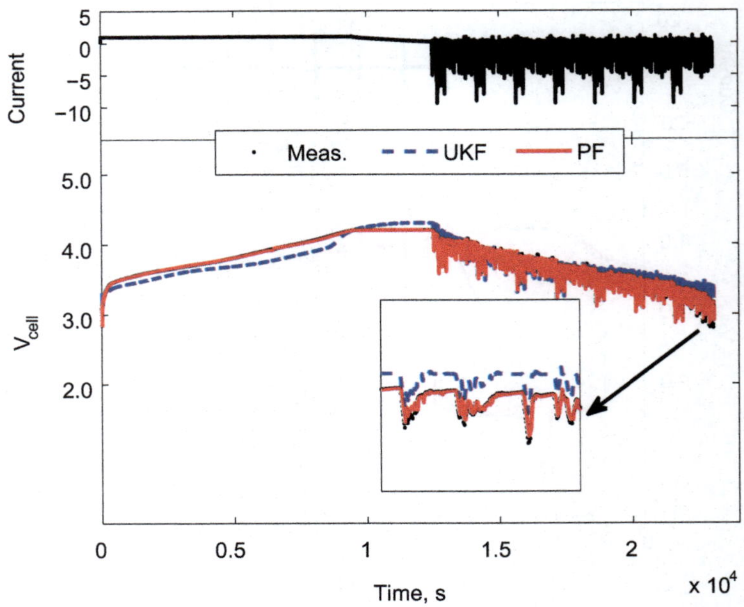

Fig. 9.3 Comparison UKF and PF for UDDS protocol

case. The UDDS profile consists of the charging at C/3 rate, while the maximum discharge is 3C. Bottom panel of the frame compares estimated cell voltage with the measurements. The comparison near end of the discharge is also shown in the inset.

From the results presented till now, we can conclude that the PF prediction accuracy is consistently higher than the UKF. We are expected to obtain higher prediction accuracy with the PF as compared to any variant of the Kalman filter. This higher accuracy is a result of the way uncertainty quantification is handled by each RBF algorithm. Kalman filter and its variants approximate the prior and posterior distributions at various levels of Gaussianity. On the contrary, PF does not make any Gaussian assumptions, thus providing higher accuracy. It is imperative here to note the difference between UKF and PF estimates. Although the UKF resolves system and observation model nonlinearity, the propagated and the posterior uncertainties are approximated using a Gaussian distribution. As the cell voltage calculation routine of the ROM is highly nonlinear, the propagated cell voltage uncertainty is a heavy-tailed, asymmetric, non-Gaussian distribution. The unresolved non-Gaussianity results in the deviation of the UKF estimates from the truth. As the PF does not have any such assumptions, resultant estimates track the truth with a better accuracy.

Although more accurate, numerical implementation of PF is computationally expensive. Here, we compare accuracy and computational cost of the PF and UKF implementations for various test cases. In Fig. 9.4, we compare performance of PF and UKF against the data sampling frequency. For this particular comparison, nonuniformly sampled cell voltage data is interpolated to obtain the measured cell voltage

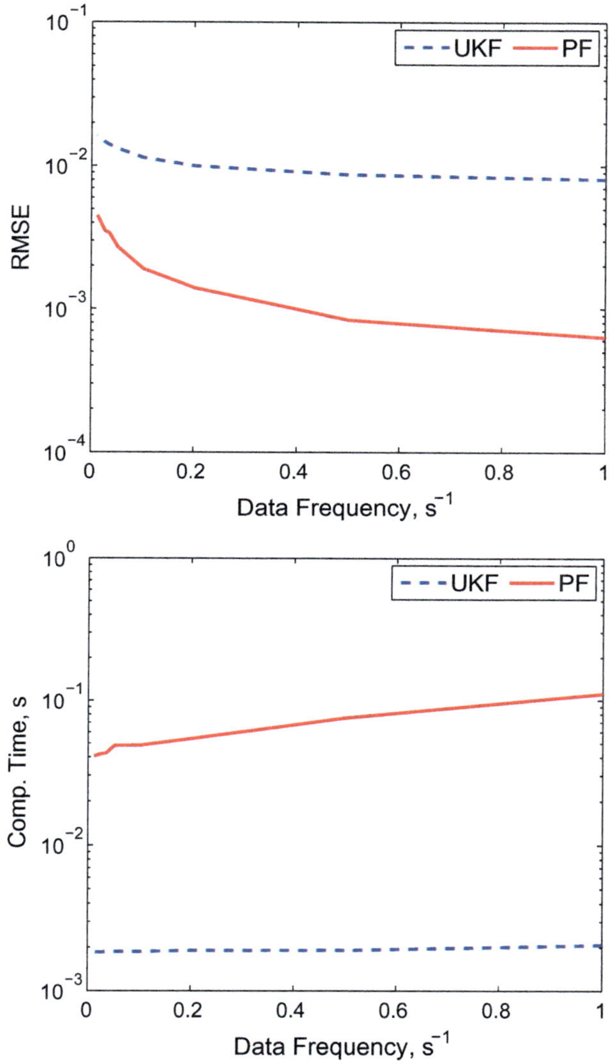

Fig. 9.4 Comparison UKF and PF algorithm

at each second. Top panel of the figure compares RMSE for UKF and PF. For all the test cases presented here, RMSE for the UKF is an order of magnitude higher than the PF. Bottom panel of the figure compares computational time for UKF and PF. The computational time for UKF execution is more than ten times lower than the PF. This result motivates a compromise between accuracy and the computational cost for selection of the filtering algorithm. UKF can be used when the computational

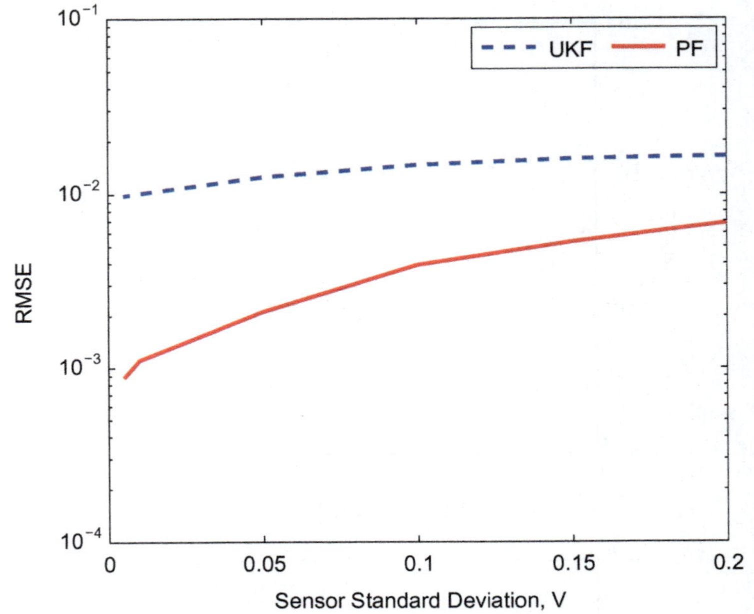

Fig. 9.5 Effect of sensor precision on UKF and PF performance

efficiency is important with the moderate accuracy requirements. However, when the accuracy is important, PF should be used.

Quality of the sensor measurements significantly impacts accuracy of the RBF predictions. In Fig. 9.5, we investigate effect of the sensor precision, specified in terms of the standard deviation, on the filter accuracy. Effect of the sensor precision is significant on the PF performance as compared to the UKF. Accuracy of the PF is higher than UKF in the overall range, while the accuracy of the PF improves significantly as compared to the UKF when high-precision sensors are used.

Results presented till this point demonstrate capability of the framework to accurately estimate the battery state. The framework assimilates the onboard measurements with the ROM predictions to significantly improve the accuracy and gives good estimates even when the ROM is not a perfect match to the measurements (typically near the end of discharge). For all the test cases presented, estimation accuracy of the PF is higher than the UKF.

Recursive Bayesian filtering algorithms are traditionally used to improve model prediction accuracy. However, as we demonstrate next, the improved prediction accuracy can be used to new physical insights from the state estimation. To investigate this capability, we consider the Li-ion state estimation at low temperature.

Figure 9.6 shows comparison of the PF cell voltage estimate with the measurements at different temperatures. Due to the higher accuracy at low temperatures, only the PF estimates are presented. For brevity, results for the C/1 discharge are presented; however, similar conclusions can be drawn for other discharge rates.

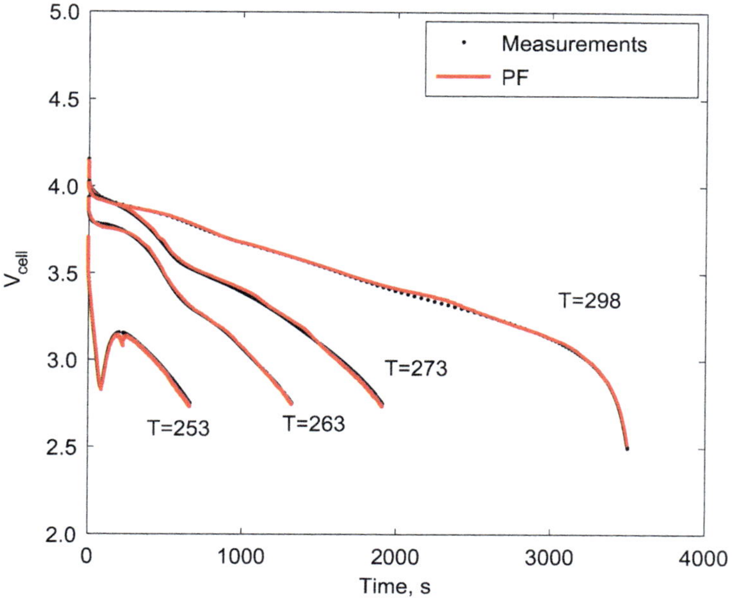

Fig. 9.6 Voltage prediction at low temperature using PF

The estimated cell voltage matches closely with the measurements at all temperatures. At temperature $T = 253K$, the cell voltage shows an uncharacteristically non-monotonous behavior that is not observed at higher temperatures. Though the physics-based model could not resolve this behavior, the PF estimate accurately tracks the non-monotonous behavior after data assimilation.

As ROM uniquely maps the cell states to the voltage, close match between estimated cell voltage and the measurements ensures that the inferred states are close to the true states. This presents an opportunity to infer hitherto unknown physics using the framework. The framework is used here to infer the cell behavior at low temperatures.

In Fig. 9.7, we compare estimated interfacial flux at different temperatures. The interfacial fluxes exhibit similar behavior for both the electrodes, as can be observed from the figure. Both the electrode fluxes are initialized from equilibrium conditions, while the flux increases consistently as the discharge current is drawn from the cell. The flux gradient decreases with the temperature, depicting the temperature dependence of the electrolyte diffusivity. For the low temperature ($T = 253K$), the flux and gradients are low and the steady-state condition is not achieved before the end of discharge. It can be seen that for all temperatures till $T = 263K$, the slope of q_{2k} is the same. At $T = 253K$, however, the slope of q_{2p} is higher than q_{2n}. This implies that the electrolyte diffusion limitations at low temperature are causing accumulation of lithium ions in the negative electrode section resulting in

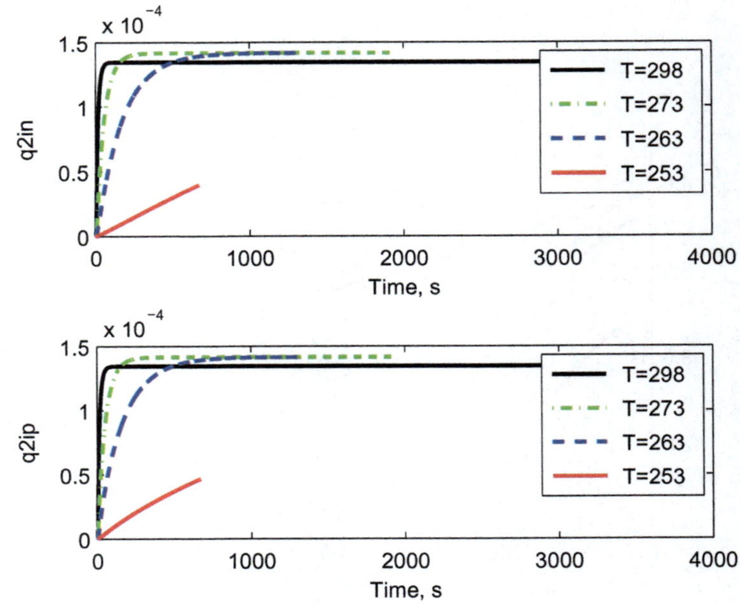

Fig. 9.7 State estimation using particle filter. This figure shows evolution of interfacial flux for negative (top) and positive (bottom) electrodes

the non-monotonic voltage response. This feature is explained further by analysis of the concentration in the electrolyte.

Figure 9.8 shows estimated ionic concentration at the interface (frame (a)) and the current collectors (frame (b)). Ionic concentration at the separator mid location is also shown in the frame (b). The figure depicts ionic flow across the cell from the negative electrode to the positive electrode current collector. For $T = 298K$ to $T = 263K$, the framework estimates a consistent ionic flow from negative to the positive electrode with the lower operating temperature delaying the steady state. At $T = 253K$, the framework estimates ion accumulation at the negative electrode interface, as can be observed by comparing frames (a) and (b). The difference between times required to reach the steady state in frames (a) and (b) is an indication of the electrolyte transport timescales. It can be seen that between (a) and (b), the difference in the times increases as operating temperature decreases.

This implies that the limiting process is the lithium transport in the electrolyte phase. During discharge, lithium reacts and de-intercalates from the negative electrode, is transported through the electrolyte, and reaches the positive electrode to be intercalated back into the positive electrode particles. At $T = 253K$, accumulation in the negative electrode implies that the effect of temperature on reaction is lesser in comparison to the effect on lithium transport. This results in the larger lithium release into the electrolyte at the negative electrode, and larger depletion of ions from the positive electrode region. But the electrolyte is unable to maintain the lithium-ion

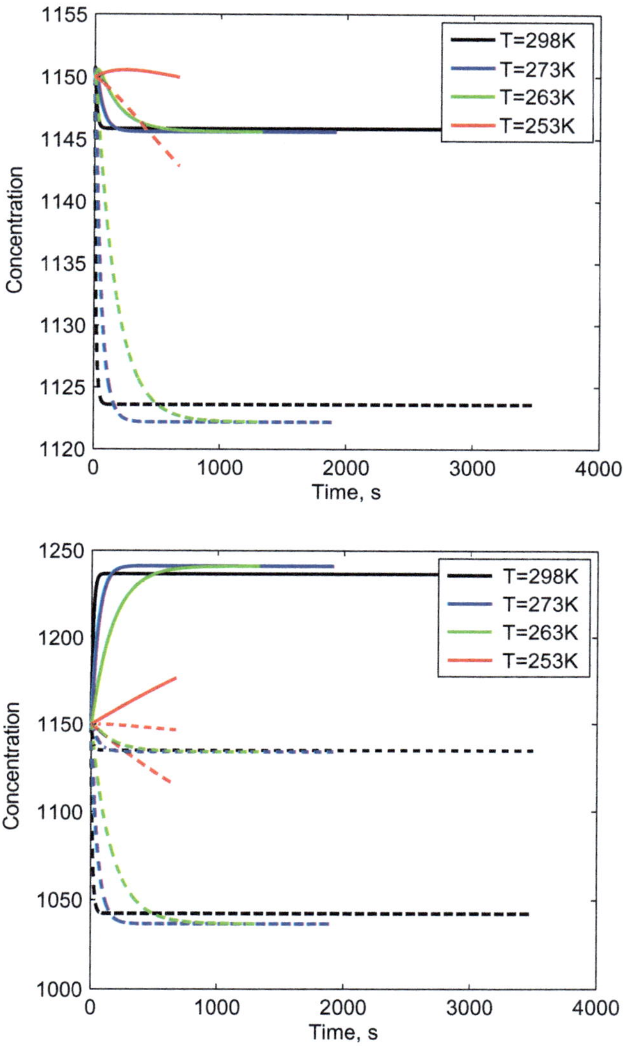

Fig. 9.8 Figure shows estimated ionic concentration at (**a**) interfaces and (**b**) current collectors. Ionic concentration at the separator mid is also shown in the frame (**b**). Ionic concentration at negative electrode is indicated using a solid line, at positive electrode using a dashed line, and at the separator mid using a dash-dot line

supply to reach a steady state, resulting in accumulation in the negative region. This accumulation could result in re-intercalation of lithium into the negative electrode and could be the cause of the non-monotonic voltage response.

Figure 9.9 shows solid-phase lithium concentration at the negative (top panel) and the positive electrodes (bottom panel). It can be seen from Fig. 9.9 that the c_{1n} decreases and c_{1p} increases as discharge proceeds at room temperature. At low tem-

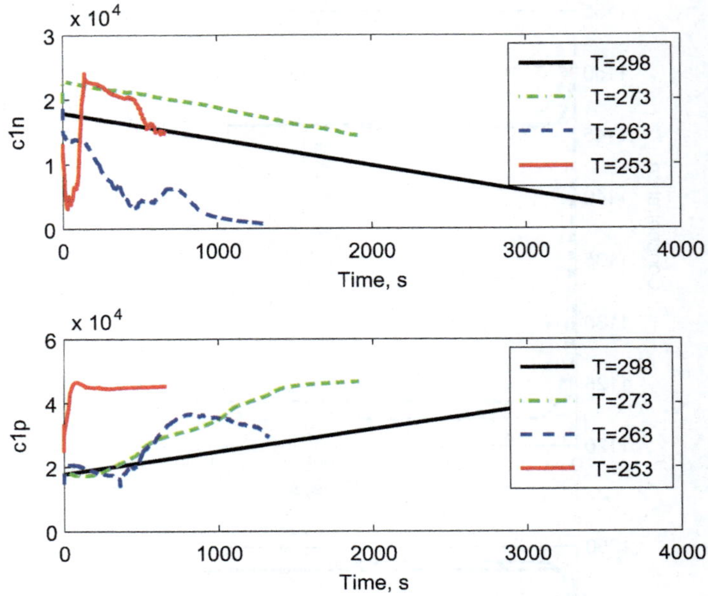

Fig. 9.9 State estimation using particle filter. This figure shows evolution of surface lithium concentration for negative (top) and positive (bottom) electrodes

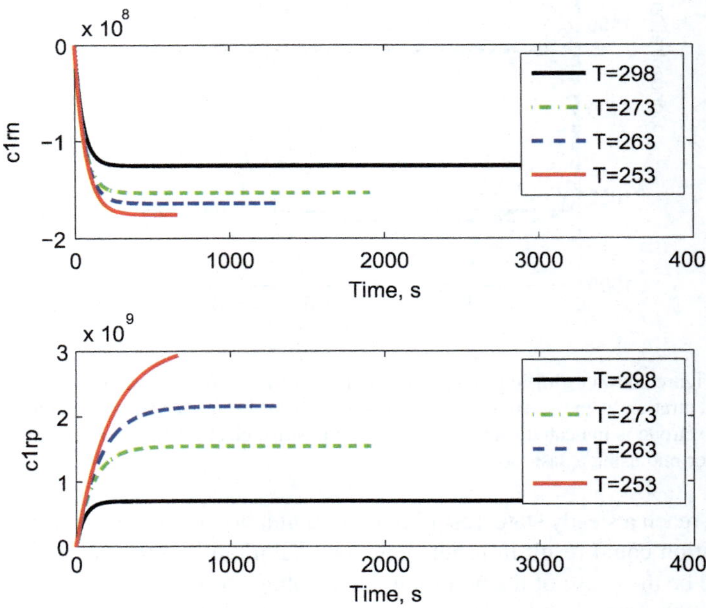

Fig. 9.10 State estimation using particle filter. This figure shows evolution of surface lithium concentration for negative (top) and positive (bottom) electrodes

perature, however, for an accurate match of cell voltage at $T = 263K$ and $T = 253K$ (Fig. 9.6), the filter intervention results in faster emptying out of the negative electrode at 263 K and filling up of the positive electrode at 253 K. At $T = 253K$, the positive electrode is estimated to saturate after around 90s of discharge, after which the lithium concentration remains constant. However, an uncharacteristically non-monotonous behavior is estimated for the negative electrode at $T = 253K$. The negative electrode empties rapidly till around 90s, after which the electrode is estimated to fill up, resulting in the self-recharge, and subsequently empties again. As the lithium ions accumulate at the negative electrode interface (Fig. 9.9a), some of the ions are estimated to intercalate back into the electrode. This causes a partial filling up of the negative electrode. This estimated negative electrode behavior results in the accurate representation of the non-monotonic cell voltage, as observed in Fig. 9.6.

The estimated solid-phase lithium concentration gradient is shown in Fig. 9.10. Absolute value of the radial concentration gradient increases with the decreasing temperature. Steeper concentration gradients indicate reduced diffusivity of lithium inside the electrodes.

From the discussion in this section on analysis of the state variables, it can be inferred that the low-temperature operation is controlled by the electrolyte mobility in the negative electrode. This results in accumulation of lithium ions in this region.

Purpose of any BMS is accurate estimation of SOC. The present model based on electrochemical framework enables SOC estimation based on the actual lithium concentration in the electrodes. This definition of SOC based on the stoichiomet-

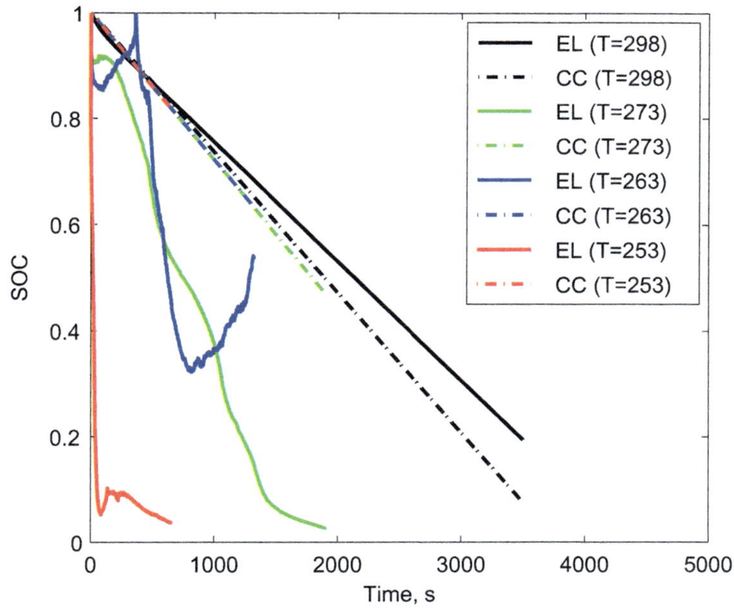

Fig. 9.11 State estimation using particle filter. This figure shows evolution of surface lithium concentration for negative (top) and positive (bottom) electrodes

ric ratio of the respective electrode is compared against the SOC estimated using Coulomb counting (CC). As the electrode-based SOC cannot be measured directly, accuracy of the SOC estimation should be inferred from the accuracy of the voltage prediction. Given that the model predicts the cell voltage very accurately across the operating temperatures and types of operations, the electrode SOC estimation can be claimed to be equally accurate. This subsection presents the relative merits of the electrode-based SOC estimation against the CC-based SOC.

Due to the higher stability, positive electrode SOC is used here for analysis. In order to match the convention that cell SOC has to decrease during discharge, the electrode SOC is defined as

$$SOC_{el}(t) = (C_{spmax} - C_{sp}(t))/C_{spmax}, \tag{9.9}$$

while the CC-based SOC is defined as

$$SOC_{cc}(t) = 1 - \frac{\int_0^t idt}{Q_{max}}. \tag{9.10}$$

Figure 9.11 shows the electrode SOC estimates at different operating temperatures. The figure also compares SOC_{cc} with the SOC_{el}. At room temperature, SOC_{cc} match closely with SOC_{el} and end near zero. At low temperature, SOC_{el} depicts non-monotonous behavior and reach near zero value. However, SOC_{cc} is a linear function for a constant current discharge and cannot capture the non-monotonicity at low temperature. Thus, SOC_{cc} wrongly predicts remnant charge at low temperature, even though the cell has reached the minimum voltage. SOC_{el} accurately captures this behavior as temperature dependencies are inherently considered in the computation. To improve prediction accuracy, definition of SOC_{cc} must consider temperature dependence of the maximum capacity Q_{max}, which is often improbable to obtain.

Figure 9.12 compares the electrode SOC with the SOC_{cc} for the HPPC (frame (a)) and UDDS (frame (b)) protocols. It can be seen from Fig. 9.12a that the electrode SOC matches with SOC_{cc} at the end points. During the rest period in HPPC (shown in inset), the current is zero and the SOC_{cc} does not change. During these periods, however, the cell voltage does change (Fig. 9.6) due to the relaxation of the electrochemical processes. This relaxation effect is captured well by the electrode SOC. In a realistic driving scenario, where continuous power demand is interposed by rest periods, the electrode SOC can give a better estimate of the actual SOC of the cell, by accounting for inherent relaxation processes.

It can be seen from Fig. 9.12b that although the SOC_{cc} is close to zero, the electrode SOC estimates a finite lithium concentration in the electrodes. In the inset, the electrode SOC is estimated from a different initial state. It can be seen that the electrode SOC trajectory is unique. This establishes that the electrode SOC is independent of the initial condition. In comparison, estimation of SOC_{cc} would require the complete current history.

Although easy to compute, SOC_{cc} is known to have dependence on the cell capacity and operation history. During an onboard scenario, both these might be difficult

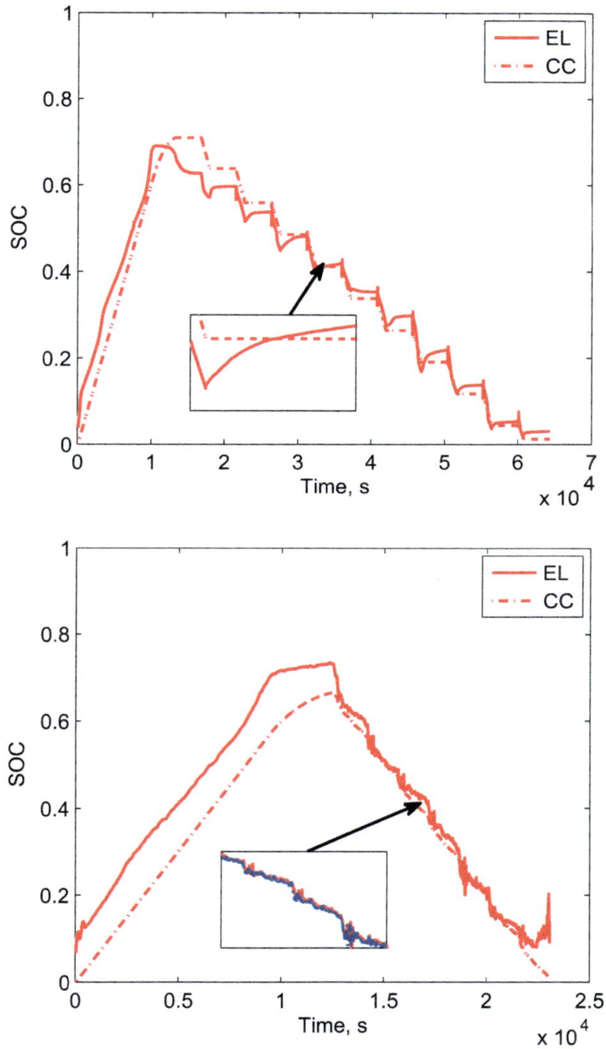

Fig. 9.12 Comparison of electrode SOC estimate with coulomb counting based SOC for (**a**) HPPC protocol; and (**b**) UDDS protocol. Inset of frame (**a**) shows comparison of two SOCs near one of the HPPC pulses. Inset of frame (**b**) compares estimated electrode SOCs initialized from different initial conditions. The blue line depicts estimated electrode SOC when the UDDS protocol is initialized from a fully charged condition

to obtain, leading to erroneous estimates. From the discussion on low-temperature results, it can be seen that the electrode SOC does not require a priori knowledge of cell capacity at the respective temperature. The UDDS results indicate that electrode SOC is independent of current history also, in addition to representing the cell relax-

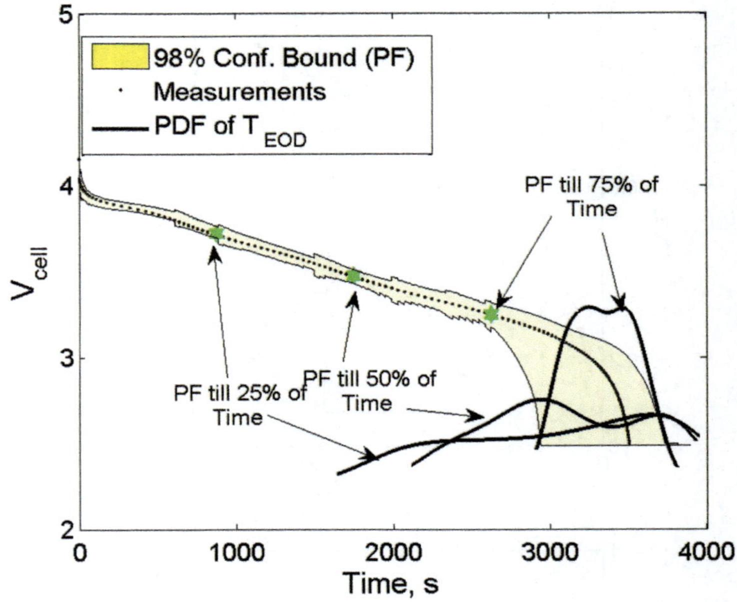

Fig. 9.13 Predictive capability of the particle filter at room temperature. Figure shows voltage and end of discharge prediction. 98% confidence bound on voltage and PDF of the end of discharge prediction is also shown

ations accurately. Given that estimation of electrode SOC requires about $0.1s$ for a second of real-time operation, it can be concluded that electrode SOC is amenable for onboard scenarios as well.

To establish predictive capability of the proposed framework, the cell voltage data till 75% of the end of discharge time is assimilated and the behavior for the remaining 25% of the time is predicted. Figure 9.13 depicts the predictive capability of the framework for $C/1$ discharge at room temperature. The figure shows a 98% confidence bound on the cell voltage predictions when the data is assimilated till 75% of the *true* end of discharge time (T_{EOD}). The measured cell voltage resides within the 98% confidence bound for the complete discharge cycle. The figure also shows PDF of T_{EOD} after the data assimilation till 25, 50, and 75% of the end of discharge. The PDF becomes narrower as more data is assimilated, signifying the increasing prediction accuracy and confidence.

9.2 State of Health Estimation Using Machine Learning

In this section, we will explore applicability of the machine learning algorithms for a battery state of health estimation. Without loss of generality, we will explore support vector machine (SVM) for classification and regression. However, other machine learning approaches can be similarly used in the same framework.

Table 9.2 List of batteries with their operating parameters

Battery number	Discharge current	End voltage	End-of-life condition	Operating temperature	No. of cycles
B0005		2.7V			168
B0006	2A Constant Current	2.5V	30% fade in rated capacity	24 C	168
B0007		2.2V	(2Ahr to 1.4Ahr)		168
B0033		2.0V			197
B0034	4A	2.2V	Capacity reduced to 20%	24 C	197
B0036	2A	2.7V	fade (1.6Ahr)		197
B0038	Multiple- 1A	2.2V			47
B0039	2A and 4A	2.5V	Capacity reduced to 20%	24C & 44C	47
B0040		2.7V	fade (1.6Ahr)		47
B0042	Multiple-	2.2V			112
B0043	1A, 4A	2.5V	Capacity reduced to 30%	4C	112
B0044		2.7V	fade (1.4Ahr)		112
B0045		2.0V			72
B0046	Fixed Load	2.2V	Capacity reduced to 30%		72
B0047	1A	2.5V	fade (1.4Ahr)	4C	72
B0048		2.7V			72
B0054	Fixed Load	2.2V	Capacity reduced to 30%		103
B0055	2A	2.5V	fade (1.4Ahr)	4C	102
B0056		2.7V			102

First step in the implementation of any machine learning algorithm is to select an appropriate dataset for training. In this work, we have sourced the battery cycling data from a publicly available repository, provided by Prognostics Center of Excellence (PCoE) at Ames Research Center, NASA (http://ti.arc.nasa.gov/tech/dash/pcoe/rognostic-data-repository/). This repository contains cycling data from 34 different batteries. These batteries are sequentially numbered prefixed by B (for e.g., B0001). Out of these batteries, the end-of-life (EOL) condition was not defined properly for 12 batteries (B0025 to B0032 and B0049 to B0052). Hence, these batteries are not used in this work. Out of the remaining 22 batteries, further pruning is done and following three batteries are eliminated: B0018, B0041, and B0053. Battery number B0018 is not considered in this study because, though its discharge current, end voltage, operating temperature, and EOL condition are same as battery number B0006, it has cycled only 132 cycles, whereas rest of batteries in its group have cycled 168 cycles. For similar reasons, batteries B0041 and B0053 are not considered. After this pruning, details of the remaining 19 batteries considered in this work are provided in Table 9.2.

We use this dataset to train a model for remaining useful life (RUL) estimation of the battery. For the RUL estimation, we combine both the classification and regression approaches. Thus, our overall RUL estimation approach follows two steps:

- Step I : Classification. The complete dataset is classified into four sets. Each class is labeled as A, B, C, and D, each class bracketing the batteries in 25% of RUL. The last set involves data from the last 25% of the cycles and hence close to the end of life.
- Step II: Regression. The RUL vs number of cycles of the last set is considered for exact estimation of RUL.

The advantage of the two-step method is that the load on regression step is minimal. A quick classification based on the extracted features can eliminate the cycles where the battery is healthy (initial 75% of the operation). Thus, the regression step can be potentially used on board for fast estimation of RUL. In the remaining part of this section, we will present various test cases demonstrating effectiveness of the RUL estimation using the machine learning approaches.

9.2.1 Case 1: Multiple Battery Data

In the first case study, effectiveness of the proposed method for RUL estimation is demonstrated for a set of three batteries (B0033, B0034, and B0036). As can be observed from Table 9.2, these batteries have same EOL condition, operating temperature, and number of cycle; however, their discharge current and end voltage are different. The purpose of this case study is to evaluate performance of the proposed method on a set of batteries that have similar operating characteristics but each battery has its own independent signature. This methodology helps in understanding the applicability of proposed method as a generic tool that can suit any of the batteries. The evaluation of accuracy of both the stages, classification and regression, is done separately; however, the overall time complexity of the approach is measured to check its applicability in real-time scenarios. For classification purpose, all the discharge cycles from three batteries are aggregated. The aggregated data constitutes 591 cycles, 70% of this data is considered for classification model training (414 cycles), and rest for model testing (177 cycles). To ensure an unbiased sample, the testing and training datasets are created using a random number generator. Training dataset consists of 101 cycles representing class A, 105 cycles representing class B, 105 cycles representing class C, and 103 cycles representing class D. Classification model is created using SVM classifier with radial basis function as the kernel. During training phase, parameters are optimized and final optimized parameter list is derived to create a robust model. Features selected for final training after optimization are VCE and VCFI. The model with the above-mentioned two parameters is tested with the 30% of the data (177 cycles). The model correctly classifies 161 out of 177 cycles; thus the resultant classification accuracy is 94.15%. The root-mean-squared error for classification is 0.2126 with Kappa statistic of 0.8794. Table 9.3 provides detailed classification accuracy for each class along with other statistical parameters and Table 9.4 provides the confusion matrix of the classification.

The regression model is built using the class D cycle data from the dataset of three batteries. This model is built using 103 (70% of data) cycle data and remaining 44 cycles are used to test the model. The regression model is built using support vector regression (SVR) code developed in MATLAB. Based on the classifica-

Table 9.3 Class-level accuracy of the classification model

Class	True positive rate	False positive rate	Precision	Recall	F Measure	ROC area
A	0.957	0.023	0.936	0.957	0.946	0.967
B	0.933	0.023	0.933	0.933	0.933	0.955
C	0.833	0.037	0.875	0.833	0.854	0.898
D	0.909	0.038	0.889	0.909	0.899	0.936
Wt. Avg.	0.91	0.03	0.909	0.91	0.909	0.94

Table 9.4 Confusion Matrix

Classified As	Class			
	A	B	C	D
CASE 1				
A	**44**	2	0	0
B	0	**42**	3	0
C	1	1	**35**	5
D	2	0	2	**40**
CASE 2				
A	**13**	0	0	0
B	0	**12**	0	0
C	1	0	**15**	1
D	0	0	0	**17**
CASE3				
A	**21**	0	0	0
B	4	**16**	1	0
C	0	3	**19**	0
D	0	0	1	**16**
CASE 4				
A	**86**	6	0	0
B	5	**84**	3	1
C	2	3	**77**	10
D	1	0	6	**85**
CASE 5				
A	**146**	12	0	4
B	13	**123**	25	1
C	4	16	**122**	20
D	4	1	22	**135**

tion exercise, VCE and VCFI are the inputs to the SVR and RUL is the output. In Fig. 9.14, the regression model prediction is compared against the training and testing data. The root-mean-squared error for the test data is 0.2420%, and thus the resultant prediction accuracy is > 99%. The results of the model compared with the data used in the training phase are shown in Fig. 9.14a, and the % error, defined as $(1 - RUL_{data}/RUL_{model}) \times 100$ is shown in Fig. 9.14b. The corresponding results for the test data are shown in Fig. 9.14c, d. In Fig. 9.14b, d, the value of the error at 95% confidence interval is marked as the horizontal lines. The root-mean-squared error (RMSE) and the values of the lower bound (LB) and upper bound (UB) for this case are given in Table 9.5. The figure shows that the model can estimate RUL with minimal error irrespective of the battery cycle. A distinct feature of the model is extremely low errors for the training set. The training and testing datasets do not have any elements in common. Under these conditions, it is to be noted that the 95% of the data points of all the three batteries predicted by the model are within a reasonable error bounds. To establish computational efficiency of the proposed method, execu-

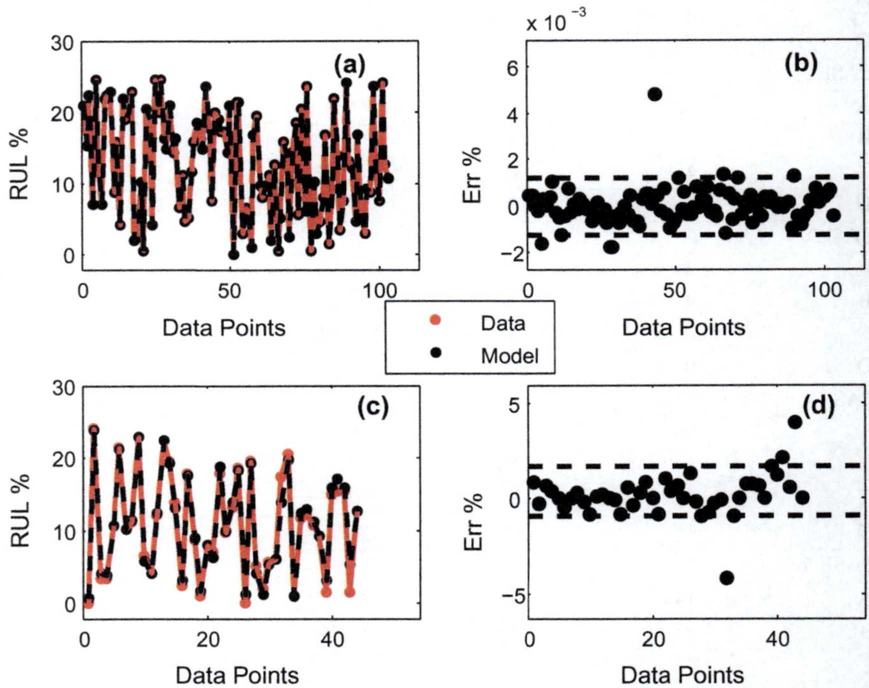

Fig. 9.14 Comparison of estimated RUL% estimated by regression model with data, for the training set that includes B0033, B0034, and B0036. **b** Error in RUL% estimation by regression model with data for the training set. The 95% confidence intervals are also shown. Frame (**c**) shows comparison of RUL% estimated by regression model with data for the testing set. Frame (**d**) shows error in RUL% estimation by regression model with data, for the testing set. The 95% confidence intervals are also shown

Table 9.5 Statistical analysis of model

	RMSE (%)	95% Lower bound	95% Lower bound
CASE 1			
Training set	6e-7	−0.0013	0.0012
Testing set	0.1659	−0.8882	1.6534
CASE 2			
Training set	2.85e-7	−0.00087	0.00112
Testing set	0.3582	−0.699	1.5007
CASE 3			
Training set	0.0004	−0.0310	0.0352
Testing set	0.8681	−7.4849	5.7270
CASE 4			
Training set	0.0005	−0.0546	0.0415
Testing set	0.3766	−7.0381	8.1358
CASE 5			
Training set	7.32e-06	−0.00495	0.00465
Testing set	0.4125	−7.986	7.1393
CASE 6			
Training set	0.0026	−0.09909	0.0729
Testing set	0.357	−10.75	7.87

tion time of the MATLAB implementation on a desktop system (Intel i3 3.2 GHz dual-core processor, 4.0 GB RAM, 32-bit Microsoft Windows 7) is investigated. The computation time for RUL estimation of a battery with 196 discharge cycles is 1.26 ms. This time includes the time taken by feature computation, classification, and regression modules.

9.2.1.1 Case 1A: RUL Prediction of an Untested Battery

To investigate accuracy of the regression model to predict RUL of a kind of a battery different from the batteries used for training, the model is trained using the class D cycle data from batteries B0033 and B0036 and tested for the battery B0034. The features selected for training are VCE and VCFI. The regression model outcome for testing and training data is shown in Fig. 9.15a–d and the detailed statistics is provided in Table 9.5. The prediction RMSE for testing data is 0.11%, making it an efficient model with average accuracy of 99.89% for RUL prediction of new kind of batteries.

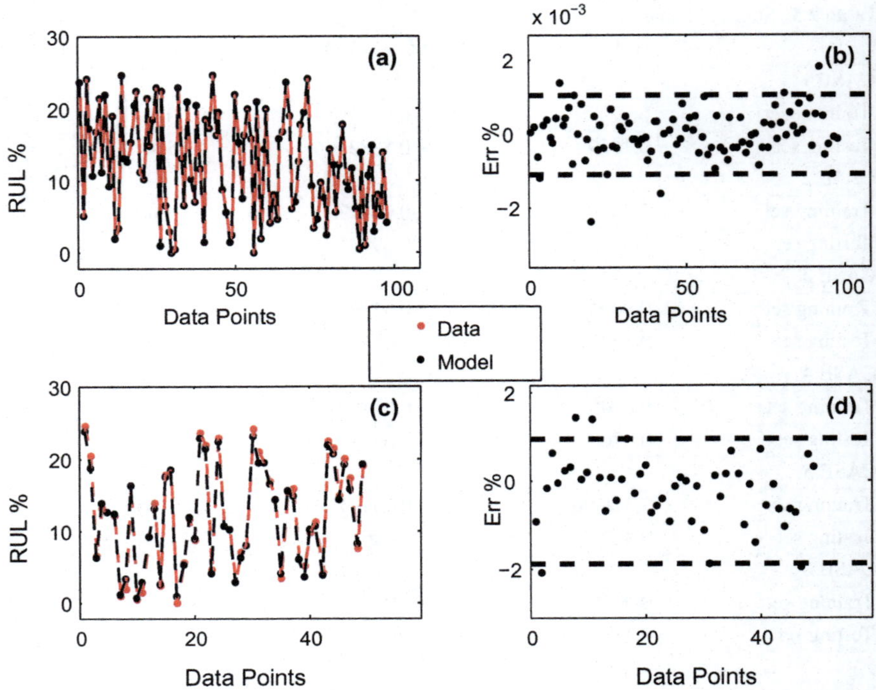

Fig. 9.15 Figure shows regression model outcome trained using data from batteries B0033 and B0036 and tested for the battery B0034. The figures a–d represent the corresponding results as in Fig. 9.14. In a, c of this figure, estimated RUL% value is represented in black color and experimental data in red color

9.2.1.2 Case 1B: Effect of SVM Classification Error

In a multistep approach, accuracy of initial steps is expected to impact the accuracy of subsequent steps. In a proposed two-step approach, misclassification in an SVM step, where a battery belonging to different classes is erroneously classified as class D, may impact RUL prediction accuracy of the SVR. To investigate the effect of SVM classification error on the RUL prediction accuracy, the regression model is trained using the class D data and tested for a randomly selected mixture of class C and class D data. For the present test case, the regression model is trained and tested for the battery B0034. Similar to the previous test cases, VCE and VCFI are used as features for training. RUL prediction accuracy of the regression model for testing and training data is shown in Fig. 9.16a–d and the detailed statistics is provided in Table 9.5. The prediction RMSE for testing data is 0.21%, while the maximum RUL prediction error is about 6%. The proposed multistep approach, thus, ensures high RUL prediction accuracy even in view of the erroneous classifier output. It should be noted that the SVM classification error primarily results in a conservative estimate

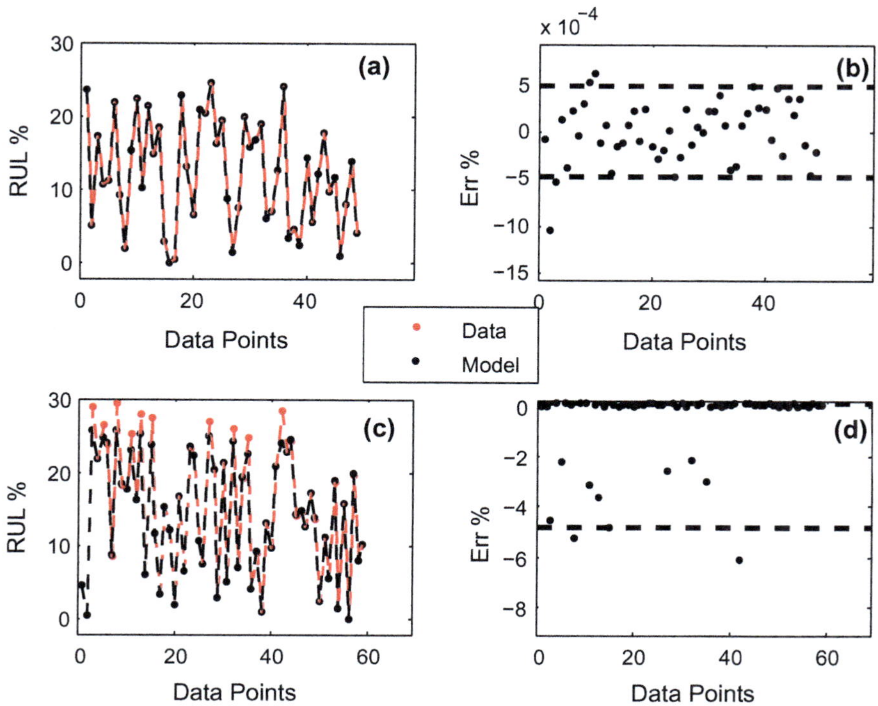

Fig. 9.16 Figure shows regression model outcome when SVM classifier erroneously classifies class C data as class D. For the test case, class D data of the battery B0034 is used for training and randomly selected 20% data from class C is used for testing. The figures a–d represent the corresponding results as Fig. 9.4. In a, c of this figure, estimated RUL% value is represented in black color and experimental data in red color

of the RUL, and thus safe battery operation is ensured even in view of the erroneous SVM classification.

9.2.2 *Prognostic Horizon of Multistep Approach*

In the proposed approach, the SVM and SVR are trained offline and subsequently used online for RUL prediction. The predicted RUL is expected to aid in the prognostic task of the BMS. In this paper, the prognostic capability of the proposed multistep approach is demonstrated using Prognostic Horizon, which is a widely used metric for comparing prognostic algorithms [46]. For this test case, class D data from batteries B0033 and B0036 is used for training the SVR and the data from battery B0034 is used for testing. The SVR is trained using VCI and VCFI as features. To account for the SVM classification error, 20% data belonging to class C is also used for testing.

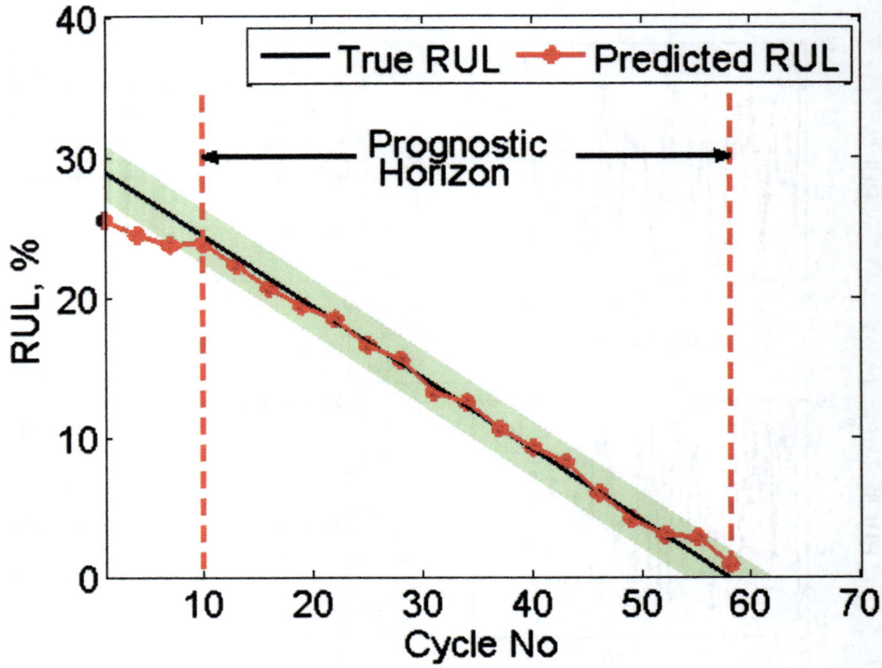

Fig. 9.17 Figure shows prognostic horizon of the proposed approach

In Fig. 9.17, the predicted RUL is compared with the true RUL. Prognostic horizon for the test case is also shown in the figure. The desired RUL prediction accuracy is given in terms of $\pm$ a bound, where a=2 is used in the present test case. The shaded region in the figure shows the desired RUL prediction accuracy. The prognostic horizon is given by the time between the algorithm first achieves a desirable accuracy to the battery end of life. The proposed algorithm has a prognostic horizon of about 50 cycles for the present test case, and accurately predicts the RUL when about 25% of the useful life of the battery is remaining.

9.2.3 Case 2: Single Battery Data

The RUL estimation model presented in this paper is tuned for a particular battery and its performance is evaluated. To create this model, complete discharge cycle data from the battery B0036 is considered. Classification model is built using 70% of 197 cycles and tested for the remaining 59 cycles. Following the parameter optimization, VCE and VCFI are identified as critical parameters and used for classification and regression model building. The classification model correctly classifies 57 out of 59 cycles, representing 96.61% accuracy and kappa of 0.9545. Table 9.4 provides the confusion matrix for this model. The corresponding regression model is built using class D cycles of battery B0036. There are total of 49 cycles belonging to class D, out of which 70% of cycles are randomly picked for training purpose and rest are used for testing the regression model using the MLP kernel. The testing outcome shows that model has very low error (RMSE) of 0.16554% while testing, making it an efficient model with average accuracy of 99.80%. Figure 9.18a–d shows the regression model outcome for training and test data and the detailed statistics is provided in Table 9.5. It can be seen that this model has very good classification as well as regression accuracy and hence can be readily used for batteries operating in same environment as B0036.

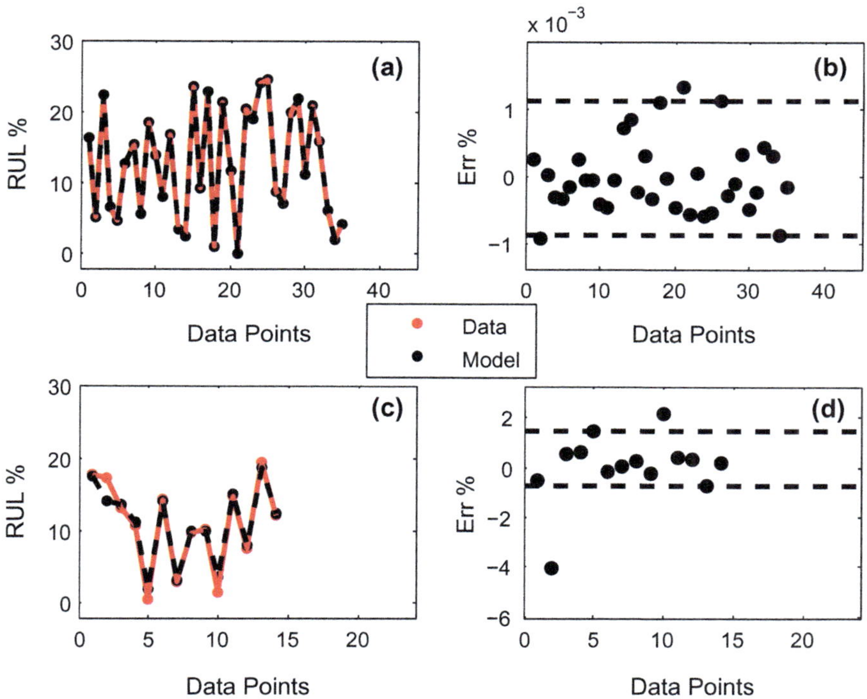

Fig. 9.18 Figure shows single battery (B0036) regression model outcome. The figures a–d represents the corresponding results as Fig. 9.14. In a, c of this figure, estimated RUL% value is represented in black color and experimental data in red color

9.2.4 Case 3: Two Batteries at Various Temperatures

The next experiment is performed on data created by combining discharge cycles of B0005 and B0056. As can be noted from Table 9.2, these batteries are tested with the same discharge parameters; however, the operating temperature of the batteries is different. B0005 is tested at an ambient temperature of 24 °C, whereas B0056 is tested at 4 °C.

Classification model is created using 70% of total discharge cycles, i.e., out of 270 cycles, 189 cycles are used for model training and remaining 81 cycles are used for model testing. 270 cycles are divided almost equally across classes; classes A and C having 68 cycles and classes B and D having 67 cycles. Model feature vectors are optimized and finally two features, VCE and VCFI, are used. The model correctly classifies 72 cycles (88.88%) with kappa of 0.8513. Table 9.4 provides the confusion matrix for this model.

Regression model is built using 47 cycles of class D data (70%). This model is built using the same input parameters defined above, VCE and VCFI. The model is tested using remaining 20 cycles of data. The testing and training datasets are mutually exclusive and are randomly selected. This model has RMSE of 0.86% on testing (results not shown), resulting in average accuracy of 99%. The 95% bounds on under- and overestimation error are given in Table 9.5.

From the results, it can be observed that model performance is low compared to single battery case. It is interesting to note that the accuracy is also lower than three battery cases discussed earlier. It is a well-known fact that batteries operated at different temperature ranges behave differently. Hence, it can be seen that the model described in this work is able to distinguish such changes due to operating conditions as well. It is also important for designer to choose right set of batteries for combining and training models.

9.2.5 Case 4, 5: Multi-battery Data at High and Low Temperatures (HT/LT)

This experiment is carried out to check performance of the proposed approach on a set of batteries operated at a similar operating temperature, but with totally different operating profiles. From Table 1, nine batteries are chosen that have operating temperature of 24 °C or 44 °C for the study at HT. For the LT studies, the data of 10 batteries at 4 °C were chosen.

To build classification model at HT, discharge cycle data from the nine batteries is aggregated, which makes total number of data point as 1236. Data splitting for testing and training is carried out in the same way as mentioned earlier. For training, total 867 cycles are used comprising 217 cycles from class A, 219 cycles from class B, 214 cycles from class C, and 217 cycles from class D. In this experiment too, SVM classifier with *Radial Bases Function* is used to build the classifier model. The features selected after parameter optimizations are VCE and VCFI. The classification accuracy for this test case is 89.97% with kappa value of 0.8663 and RMSE of 0.2239. Table 9.4 provides confusion matrix for the tested cycles. Regression model is built using 216 discharge cycles belonging to class D and rest 93 (30%) cycles are used to test the model. Model is built using the same two features as defined above, and MLP kernel is used. Figure 9.19 shows comparison of the model with respect to experimental data, and the accuracy values are given in Table 9.5. The regression model for HT scenario has a RMSE of 0.3766. The 95% over- and underestimation errors are at 8.13 and 7.04 percentages. As mentioned earlier, the accuracy is much higher for the training set. In a similar manner, the model is tested at LT case. For this experiment, all the batteries that operate with an ambient temperature of 4 °C are chosen. After classification the total dataset is 231 cycles. Regression model is built using 162 of these data, and the model is tested using 69 discharge cycles belonging to class D. Model is built using the same parameters and kernel as discussed above. For the LT case, the accuracy (Table 9.5) is in the same range as the HT, demonstrating that the model can be used with equal confidence across temperatures.

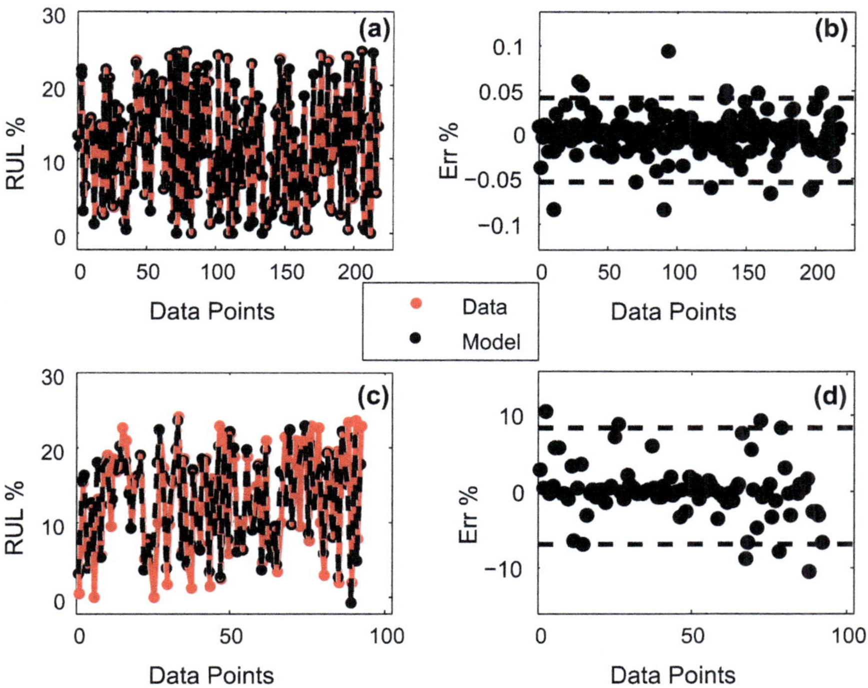

Fig. 9.19 Graph displaying regression outcome for multi-battery same temperature case (HT). In a, c the estimated RUL% value is represented in black color and experimental data in red color

9.2.6 *Case 6: All Battery Data*

Purpose of this experiment is to check generalizability of the model on a given set of diverse data. Hence, to carry out this experiment, discharge cycle data from all 19 batteries is aggregated. Total data available after aggregation is 2166 cycles. To build the classification model, the complete data is split into training (70%) and testing (30%) data. Training dataset of 1518 cycle is composed of 381 cycles from class A, 382 cycles representing class B, 377 cycles representing class C, and 378 cycles representing class D. Testing data (648 cycles) has equal representation of 162 cycles from each class. In this experiment, also same classifier parameters are used as described in previous experiments. The critical features selected for final training after optimization are VCE and VCFI. The classification model can correctly classify 526 cycles out of 648 cycles—the accuracy of classification is 81.17%. RMSE of this model is 0.3068 with better Kappa value of 0.749. Table 9.4 provides confusion matrix.

To build regression model, 540 discharge cycles belonging to class D are chosen. 70% of this data (378 cycles) is used for regression model training and remaining 30% of data (162 cycles) is used for testing the model. Model is built using the

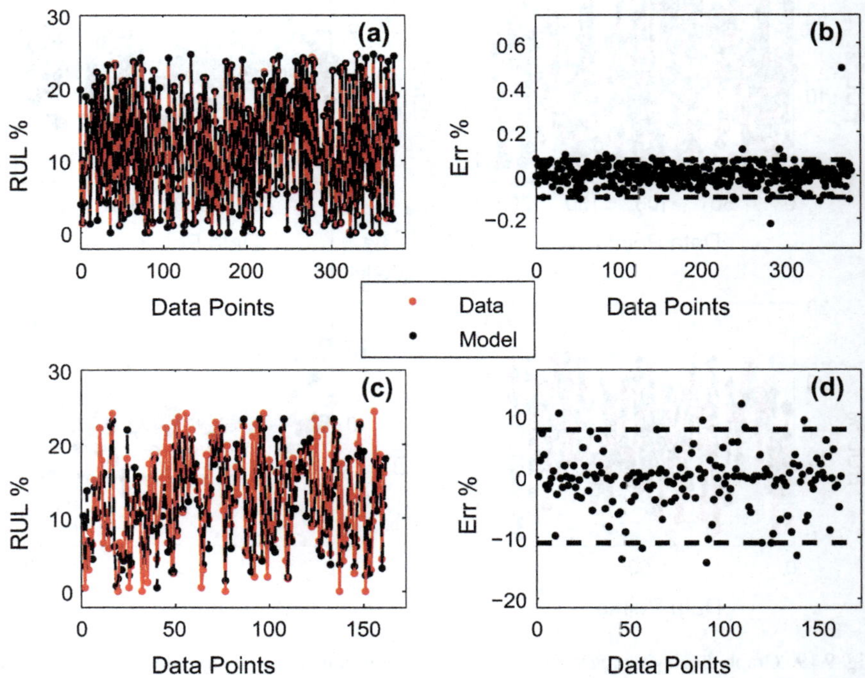

Fig. 9.20 Graph displaying outcome of all battery regression model. In a, c the estimated RUL% value is represented in black color and experimental data in red color

above-mentioned features and SVR with MLP kernel is used to build the regression model. The model performance against the experimental data is shown in Fig. 9.20.

This model has a RMSE of 0.357%, with 95% over- and underestimation errors of 7.87 and 10.75% (Table 9.5). The models built in this experiment have larger errors compared to the ones discussed in previous scenarios. This level of performance is justifiable as the input data is diverse in all directions of measurement. The average accuracy of both classification and regression models although is comparable with existing literature; it should be noted that this experiment involves multi-battery data. Hence, this kind of generic models can be used for the batteries whose past operating history is unknown. It is also interesting to observe that cycling data of multiple batteries can be represented in terms of only two critical feature vectors. The accuracy can be enhanced further by training the model with larger datasets.

9.3 Summary

This part deals with developing data-driven methods for state estimation. These state estimators integrate ROM with the onboard sensor measurements for SOC and SOH estimations. We have explored a recursive Bayesian filtering framework for state of charge estimation. This framework can be numerically implemented using a filtering method like Kalman filter, unscented Kalman filter, particle filter, etc. at a varied level of accuracy. In particular, we have compared unscented Kalman filter and particle filter for SOC estimation. When computational cost is not a constraint and higher accuracy is required, particle filter is an excellent choice for the SOC estimation. However, when computational cost is a constraint, UKF provides desired accuracy with computational efficiency. This framework can also be used for inferring hitherto unknown physics, as is demonstrated for a low-temperature operation.

We also demonstrated the use of machine learning for state of health estimation. A novel data-driven method for Li-ion battery RUL estimation is presented in this work. The proposed approach uses a combination of classification and regression stages to develop an efficient RUL estimation algorithm which is amenable for onboard implementation. Effectiveness of the algorithm is demonstrated using a publicly available cell cycling data. A minimal set of critical features is extracted from the battery dataset and is used as the input to the classification and regression models. The multistage method, in addition to increasing the accuracy and reduction in simulation time, also enables accurate RUL estimation of multiple batteries simultaneously and hence is generic in nature. The model is tested for various case scenarios, for batteries at varying temperatures and operating conditions with sufficient accuracy. From the model performance for the case studies presented in the earlier sections, it can be concluded that this approach can be used as a potential onboard RUL estimation tool for EVs. Once the trained model is used online, from measured signals, the critical features can be estimated. Based on these critical features, a quick classification step can identify if the battery is near EOL. If that were the case, the regression model can subsequently estimate the accurate RUL, and this information can be indicated to the driver while the vehicle is in operation.

References

1. M.S. Arulampalam, S. Maskell, N. Gordon, T. Clapp, A tutorial on particle filters for online nonlinear/non-gaussian bayesian tracking. IEEE Trans. Signal Process. **50**(2), 174–188 (2002)
2. G. Bai, P. Wang, H. Chao, M. Pecht, A generic model-free approach for lithium-ion battery health management. Appl. Energy **135**, 247–260 (2014)
3. T.M. Bandhauer, S. Garimella, T.F. Fuller, A critical review of thermal issues in lithium-ion batteries. J. Electrochem. Soc. **158**, R1–R25 (2011)
4. D. Bernardi, E. Pawlikowski, J. Newman, A general energy balance for battery systems. J. Electrochem. Soc. **132**(1), 5–12 (1985)
5. M. Broussely, S. Herreyre, P. Biensan, P. Kasztejna, K. Nechev, R.J. Staniewicz, Aging mechanism in li ion cells and calendar life predictions. J. Power Sources **97**, 13–21 (2001)
6. D. Brown, R.G. Landers, Control oriented thermal modeling of lithium ion batteries from a first principle model via model reduction by the global arnoldi algorithm. J. Electrochem. Soc. **159**(12), A2043–A2052 (2012)
7. L. Cai, R.E. White, An efficient electrochemical-thermal model for a lithium-ion cell by using the proper orthogonal decomposition method batteries and energy storage. J. Electrochem. Soc. **157**(11), A1188–A1195 (2010)
8. L. Cai, R.E. White, Mathematical modeling of a lithium ion battery with thermal effects in COMSOL Inc., Multiphysics (MP) software. J. Power Sources **196**(14), 5985–5989 (2011)
9. R. Darling, J. Newman, Modeling side reactions in composite li_y mn_2 o_4 electrodes. J. Electrochem. Soc. **145**(3), 990–997 (1998)
10. M. Doyle, T.F. Fuller, J. Newman, Modeling of galvanostatic charge and discharge of the lithium/polymer/insertion cell. J. Electrochem. Soc. **140**(6), 1526–1533 (1993)
11. M. Doyle, J. Newman, A.S. Gozdz, C.N. Schmutz, J.Ă.Ř.M. Tarascon, Comparison of Modeling Predictions with Experimental Data from Plastic Lithium Ion Cells. J. Electrochem. Soc. **143**(6), 1890–1903 (1996)
12. D. Fink, A compendium of conjugate priors (1997), p. 46, http://www.people.cornell.edu/pages/df36/CONJINTRnew%20TEX.pdf
13. C. Forgez, D.V. Do, G. Friedrich, M. Morcrette, C. Delacourt, Thermal modeling of a cylindrical LiFePO4/graphite lithium-ion battery. J. Power Sources **195**(9), 2961–2968 (2010)
14. T.F. Fuller, M. Doyle, J. Newman, Simulation and optimization of the dual lithium ion insertion cell. J. Electrochem. Soc. **141**(1), 1–10 (1994)
15. P. Gambhire, K.S. Hariharan, A. Khandelwal, S.M. Kolake, T. Yeo, S. Doo, A physics based reduced order aging model for lithium-ion cells with phase change. J. Power Sources **270**, 281–291 (2014)
16. P. Gambhire, N. Ganesan, S. Basu, K.S. Hariharan, S.M. Kolake, T. Song, D. Oh, T. Yeo, S. Doo, A reduced order electrochemical thermal model for lithium ion cells. J. Power Sources **290**, 87–101 (2015)

K. S. Hariharan et al., *Mathematical Modeling of Lithium Batteries*,
Green Energy and Technology, https://doi.org/10.1007/978-3-319-03527-7

17. R.E. Gerver, J.P. Meyers, Three-dimensional modeling of electrochemical performance and heat generation of lithium-ion batteries in tabbed planar configurations. J. Electrochem. Soc. **158**(7), A835–A843 (2011)
18. N.J. Gordon, D.J. Salmond, A.F.M. Smith, Novel approach to nonlinear/non-gaussian bayesian state estimation. IEE Proc. F **140**(2), 107–113 (1993)
19. W.B. Gu, C.Y. Wang, Thermal-electrochemical modeling of battery systems. J. Electrochem. Soc. **147**(8), 2910–2922 (2000)
20. M. Guo, G. Sikha, R.E. White, Single-particle model for a lithium-ion cell: thermal behavior. J. Electrochem. Soc. **158**(2), A122–A132 (2011)
21. K.S. Hariharan, A coupled nonlinear equivalent circuit thermal model for lithium ion cells. J. Power Sources **227**, 171–176 (2013)
22. F. Jiang, P. Peng, Y. Sun, Thermal analyses of LiFePO4/graphite battery discharge processes. J. Power Sources **243**, 181–194 (2013)
23. S. Julier, J. Uhlman, H.F. Durant-Whyte, A new method for the non-linear transformation of means and covariances in filters and estimators. IEEE Trans. Autom. Control **45**, 477–482 (2000)
24. R.E. Kalman, A new approach to linear filtering and prediction problem. Trans. ASME-J. Basic Eng. 35–45 (1960)
25. L.W. Kang, X. Zhao, J. Ma, A new neural network model for the state-of-charge estimation in the battery degradation process. Appl. Energy **121**, 20–27 (2014)
26. A. Khandelwal, K.S. Hariharan, V.S. Kumar, P. Gambhire, S.M. Kolake, D. Oh, S. Doo, Generalized moving boundary model for charge-discharge of LiFePO4/C cells. J. Power Sources **248**, 101–114 (2014)
27. A. Khandelwal, K.S. Hariharan, P. Gambhire, S.M. Kolake, T. Yeo, S. Doo, Thermally coupled moving boundary model for charge-discharge of LiFePO4/C cells. J. Power Sources **279**, 180–196 (2015)
28. K. Kumaresan, G. Sikha, R.E. White, Thermal model for a Li-ion cell. J. Electrochem. Soc. **155**(2), A164–A171 (2008)
29. A.H.D. McCleary, J.P. Meyers, B. Kim, Three-dimensional modeling of electrochemical performance and heat generation of spirally and prismatically wound Lithium-ion batteries. J. Electrochem. Soc. **160**(11), A1931–A1943 (2013)
30. R.N. Likelihood, J. Am. Stat. Assoc. **95**, 1335–1340 (2000)
31. J. Newman, *Electrochemical Systems* (Prentice Hall, Englewood Cliffs, NJ, 1991)
32. J. Newman, Thermoelectric effects in electrochemical systems. Ind. Eng. Chem. Res. **34**(10), 3208–3216 (1995)
33. J. Newman, W. Tiedemann, Porous-electrode theory with battery applications. AIChE J. **21**(1), 25–41 (1975)
34. C.R. Pals, J. Newman, Thermal modeling of the lithium/polymer battery: I. discharge behavior of a single cell. J. Electrochem. Soc. **142**(10), 3274–3281 (1995)
35. A.M. Patil, P. Tagade, K.S. Hariharan, S.M. Kolake, T. Song, T. Yeo, S. Doo, A novel multistage support vector machine based approach for Li ion battery remaining useful life estimation. Appl. Energy **159**, 285–297 (2015)
36. E. Prada, D.D. Domenico, Y. Creff, J. Bernard, V. Sauvant-Moynot, F. Huet, Simplified electrochemical and thermal model of lifepo4-graphite li-ion batteries for fast charge applications. J. Electrochem. Soc. **159**(9), A1508–A1519 (2012)
37. R. Paulo, Default priors for gaussian processes. Ann. Stat. **33**, 556–582 (2005)
38. L. Rao, J. Newman, Heat-generation rate and general energy balance for insertion battery systems. J. Electrochem. Soc. **144**(8), 2697–2704 (1997)
39. R.E. Kass, B.P. Carlin, A. Gelman, R.M. Neal, Markov Chain Monte Carlo in practice: a roundtable discussion. Am. Stat. **52**, 93–100 (1998)
40. J. Remmlinger, S. Tippmann, M. Buchholz, K. Dietmayer, Low-temperature charging of lithium-ion cells part ii: model reduction and application. J. Power Sources **254**, 268–276 (2014)

41. S.M. Rezvanizaniani, Z. Liu, Y. Chen, J. Lee, Review and recent advances in battery health monitoring and prognostics technologies for electric vehicle (ev) safety and mobility. J. Power Sources **256**, 110–124 (2014)
42. M. Safari, C. Delacourt, Aging of a commercial graphite/LiFePO4 cell. J. Electrochem. Soc. **158**(10), A1123–A1135 (2011)
43. M. Safari, C. Delacourt, Mathematical modeling of lithium iron phosphate electrode: galvanostatic charge/discharge and path dependence. J. Electrochem. Soc. **158**(2), A63–A73 (2011)
44. M. Safari, C. Delacourt, Modeling of a commercial graphite/LiFePO4 cell. J. Electrochem. Soc. **158**(5), A562–A571 (2011)
45. A. Saxena, J. Celaya, B. Saha, S. Saha, K. Goebel, On applying the prognostic performance metrics (2009)
46. V.S. Kumar, Reduced order model for a lithium ion cell with uniform reaction rate approximation. J. Power Sources **222**, 426–441 (2012)
47. V.S. Kumar, P. Gambhire, K.S. Hariharan, A. Khandelwal, S.M. Kolake, D. Oh, S. Doo, An explicit algebraic reduced order algorithm for lithium ion cell voltage prediction. J. Power Sources **248**, 383–387 (2014)
48. K. Smith, C.-Y. Wang, Solid-state diffusion limitations on pulse operation of a lithium ion cell for hybrid electric vehicles. J. Power Sources **161**(1), 628–639 (2006)
49. V. Srinivasan, J. Newman, Existence of path-dependence in the LiFePO4 electrode. Electrochem. Solid-State Lett. **9**(3), A110–A114 (2006)
50. V. Srinivasan, C.Y. Wang, Analysis of electrochemical and thermal behavior of Li-ion cells. J. Electrochem. Soc. **150**(1), A98–A106 (2003)
51. P. Tagade, K.S. Hariharan, P. Gambhire, S.M. Kolake, T. Song, D. Oh, T. Yeo, S. Doo, Recursive bayesian filtering framework for lithium-ion cell state estimation. J. Power Sources **306**, 274–288 (2016)
52. J.-M. Tarascon, M. Armand, Issues and challenges facing rechargeable lithium batteries. Nature **414**(6861), 359–367 (2001)
53. K.E. Thomas, C. Bogatu, J. Newman, Measurement of the Entropy of Reaction as a Function of State of Charge in Doped and Undoped Lithium Manganese Oxide. Journal of The Electrochemical Society **148**(6), A570–A575 (2001)
54. L.O. Valøen, J.N. Reimers, Transport properties of LiPF6-based Li-ion battery electrolytes. J. Electrochem. Soc. **152**(5), A882–A891 (2005)
55. V. Vapnik, *The Nature Of Statistical Learning Theory* (Springer, 2013)
56. W. Waag, C. Fleischer, D.U. Sauer, Critical review of the methods for monitoring of lithium-ion batteries in electric and hybrid vehicles. J. Power Sources **258**, 321–339 (2014)
57. Y. Ye, L.H. Saw, Y. Shi, K. Somasundaram, A.A.O. Tay, Effect of thermal contact resistances on fast charging of large format lithium ion batteries. Electrochim. Acta **134**, 327–337 (2014)
58. Y. Ye, Y. Shi, A.A.O. Tay, Electro-thermal cycle life model for lithium iron phosphate battery. J. Power Sources **217**, 509–518 (2012)

Index

A
Activation function, 154
Activity coefficient, 23
Adiabatic, 114
Ambient temperature, 50
Artificial Neural Network, 154

B
Backpropagation, 157
Basis functions, 162
Battery management system, 130, 187
Bayesian framework, 130
Bayesian inference, 146
Bayes theorem, 132
Belief network, 171
Binary electrolyte, 23
Boltzmann distribution, 166
Bruggmann factor, 21
Butler–Volmer equation, 29

C
Cell voltage, 177
Charge conservation, 22
Charge flux, 16
Charge transfer reaction, 26
Classification, 146, 190
Class label, 148
COMSOL, 111
Concentrated solution theory, 17
Concentration gradient, 184
Conditional distribution, 161
Confusion matrix, 192
Contrastive-divergence, 169
Coulomb counting, 187
Cross-entropy, 149

D
Data assimilation, 130, 177, 182, 190
Data-driven, 146
Deep belief network, 166, 170
Deep learning, 166
Deep networks, 166
De-intercalation, 183
Delta function, 144
Diffusivity, 182, 185
Distributed heat source, 35

E
Electrode SOC, 187
Electronic conductivity, 111
End of discharge, 106, 177, 190
End of life, 191
Endothermic, 112
Energy, 166
Energy balance, 33
Epoch, 150
Euler solver, 177
Exothermic, 112
Extended Kalman filter, 140

G
Gaussian distribution, 177
Gaussian process, 160, 163
Gaussian process emulator, 160
Generalized moving boundary model, 42
Gibbs-Helmholtz equation, 33
Gradient, 150
Gradient descent algorithm, 150
Graphite, 47

K. S. Hariharan et al., *Mathematical Modeling of Lithium Batteries*,
Green Energy and Technology, https://doi.org/10.1007/978-3-319-03527-7

H
Heat sources, 49
Heat transfer coefficient, 112, 116
Hidden layer, 155
Hidden Markov model, 135
Hybrid pulse power characterization, 108, 177
Hyperbolic tan, 154
Hyper-parameters, 163

I
Importance sampling, 144
Intercalation, 183
Isothermal, 114

J
Jacobian, 140
Jensen's inequality, 172

K
Kalman filter, 138, 179
Kalman gain, 138

L
Learning rate, 150
LiFePO4, 47
Likelihood, 132
Likelihood function, 147
Limiting process, 109, 183
Lithium concentration, 184
Load current, 177
Logistic regression, 155
Log-likelihood, 147
Lumped model, 35

M
Machine learning, 146, 190
Marginals, 161
Markov Chain Monte Carlo, 169
Markov process, 135
Mass conservation, 18
Mass flux, 16
Maximum a-posteriory estimate, 147
Maximum likelihood estimate, 147
Mini-batch, 151, 156
Momentum, 150
Monte Carlo, 144
Multivariate Gaussian distribution, 160, 161

N
Nernst equation, 26, 27
Neuron, 154
Nickel cobalt aluminium oxide, 106
Non-informative prior, 147

O
Ohmic heat, 111
Ohmic potential drop, 109
Onsager-DeGroot-Callen method, 36

P
Parameter estimation, 146
Particle filter, 144, 177
Partition function, 166
Peltier coefficient, 36
Phase change, 99
Phase change electrodes, 118
Phase transition, 41
Porous electrode, 13
Posterior, 132, 179
Prediction, 162
Predictive probability, 132
Prior, 132, 179
Profile approximations, 67
Prognostic horizon, 197
Pseudo-2D model, 15, 111

R
Radial basis function, 192
Random function, 160, 162
Rectified linear, 154
Recursive Bayesian framework, 135, 175
Reduced order model, 65, 105, 130, 175
Regression, 146, 190, 194
Remaining useful life, 191, 192, 195, 197
Restricted Boltzmann machine, 166
Root-mean-squared error, 178, 194, 195

S
Sampling importance resampling, 144
SEI film, 119
Sensitivity analysis, 111
Shallow network, 166
Shannon's entropy, 172
Shrinking core, 99
Sigma vectors, 141
Sigmoid, 154
Softmax, 149
State of charge, 187
State of health, 146, 190

Stochastic gradient descent, 151
Stochastic process, 135
Supervised learning, 147
Support vector machine, 158, 190, 197
Support vector regression, 192, 197

T
Tafel kinetics, 97
Taylor's series expansion, 140
Thermal electrochemical model, 40
Thermal runaway, 45
Transference number, 23

U
Unscented Kalman filter, 141, 177
Unscented transform, 141
Unsupervised learning, 167
Urban dynamometer driving schedule, 109, 177

V
Velocity vector, 150
Visible layer, 155
Volume averaging, 63

Printed by Printforce, the Netherlands